D0287479

AN INTRODUCTION TO
MATHEMATICAL
MODELING

EDWARD A. BENDER

DOVER PUBLICATIONS, INC.
Mineola, New York

Bibliographical Note

This Dover edition, first published in 2000, is an unabridged reprint of
the first edition of *An Introduction to Mathematical Modeling,* published
by John Wiley & Sons as a Wiley-Interscience Publication in 1978.

Library of Congress Cataloging-in-Publication Data

Bender, Edward A., 1942–
    An introduction to mathematical modeling / Edward A. Bender.
        p. cm.
    Originally published: New York : Wiley, c1978.
    Includes bibliographical references and index.
    ISBN 0-486-41180-X (pbk.)
        1. Mathematical models. I. Title.

QA401 .B453 2000
511'.8—dc21

99-054517

Manufactured in the United States of America
Dover Publications, Inc., 31 East 2nd Street, Mineola, N.Y. 11501

PREFACE

This book is designed to teach students how to apply mathematics by formulating, analyzing, and criticizing models. It is intended as a first course in applied mathematics for use primarily at an upper division or beginning graduate level. Some course suggestions are given near the end of the preface.

The first part of the book requires only elementary calculus and, in one chapter, basic probability theory. A brief introduction to probability is given in the Appendix. In Part II somewhat more sophisticated mathematics is used.

Although the level of mathematics required is not high, this is not an easy text: Setting up and manipulating models requires thought, effort, and usually discussion—purely mechanical approaches usually end in failure. Since I firmly believe in learning by doing, all the problems require that the student create and study models. Consequently, there are no trivial problems in the text and few very easy ones. Often problems have no single best answer, because different models can illuminate different facets of a problem. Discussion of homework in class by the students is an integral part of the learning process; in fact, my classes have spent about half the time discussing homework. I have also encouraged (or insisted) that homework be done by students working in groups of three or four. We have usually devoted one class period to a single model, both those worked out in the text and those given as problems. I have also required students to report on a model of their own choosing, the amount of originality required depending on the level of the student.

Except for Chapter 6, each section of the text deals with the application of a particular mathematical technique to a range of problems. This lets the students focus more on the modeling. My students and I have enjoyed the variety provided by frequent shifts from one scientific discipline to another. This structure also makes it possible for the teacher to rearrange and delete material as desired; however, Chapter 1 and Section 2.1 should be studied first. Chapter 1 provides a conceptual and philosophical framework. The discussions and problems in Section 2.1 were selected to get students started in mathematical modeling.

Most of the material in this book describes other people's models, frequently arranged or modified to fit the framework of the text, but hopefully without doing violence to the original intentions of the model. I believe all the models deal with questions of real interest: There are no "fake" models created purely to illustrate a mathematical idea, and there are no models that have been so sanitized that they have lost contact with the complexities of the real world. Since I've selected the models, they reflect my interests and knowledge. For this I make no apology—*caveat emptor.*

The models have been chosen to be brief and to keep scientific background at a minimum. While this makes for a more lively and accessible text, it may give the impression that modeling can be done without scientific training and that modeling never leads to involved studies. I thought seriously about counteracting this by adding a few chapters, each one devoted to a specific model. Unable to find a way to do this without sacrificing "learning by doing," I abandoned the idea.

Course suggestions. On an undergraduate level, the text can be used at a leisurely pace to fill an entire year. It may be necessary to teach some probability theory for Chapter 5, and you may wish to drop Chapter 10. More variety can be obtained by using the text for part of a year and then spending some time on an in-depth study of some additional models— with guest lecturers from the appropriate scientific disciplines if possible. Another alternative is to spend more time on simulation models after Section 5.2 if a computer is available for groups of students to develop their own in-depth models.

Acknowledgments. Particular thanks are due to Norman Herzberg for his many suggestions on the entire manuscript. My students have been invaluable in pointing out discussions and problems that were too muddled or terse to understand. I owe thanks to a variety of people who have commented on parts of the manuscript, suggested models, and explained ideas to me.

I'd appreciate hearing about any errors, difficulties encountered, suggestions for additional material, or anything else that might improve future editions of this book.

EDWARD A. BENDER

La Jolla, California
August 1977

CONTENTS

AN INTRODUCTION TO
MATHEMATICAL
MODELING

CHAPTER 1

WHAT IS MODELING

1.1 MODELS AND REALITY

The theoretical and scientific study of a situation centers around a *model*, that is, something that mimics relevant features of the situation being studied. For example, a road map, a geological map, and a plant collection are all models that mimic different aspects of a portion of the earth's surface.

The ultimate test of a model is how well it performs when it is applied to the problems it was designed to handle. (You cannot reasonably criticize a geological map if a major highway is not marked on it; however, this would be a serious deficiency in a road map.) When a model is used, it may lead to incorrect predictions. The model is often modified, frequently discarded, and sometimes used anyway because it is better than nothing. This is the way science develops.

Here we are concerned exclusively with *mathematical models*, that is, models that mimic reality by using the language of mathematics. Whenever we use "model" without a modifier, we mean "mathematical model." What makes mathematical models useful? If we "speak in mathematics," then

1. We must formulate our ideas precisely and so are less likely to let implicit assumptions slip by.
2. We have a concise "language" which encourages manipulation.
3. We have a large number of potentially useful theorems available.
4. We have high speed computers available for carrying out calculations.

There is a trade-off between items 3 and 4: Theory is useful for drawing general conclusions from simple models, and computers are useful for drawing specific conclusions from complicated models. Since the thought habits needed in formulating models are quite similar in the two cases, it matters little what sort of models we use; consequently, I have felt free to neglect computer based models purely for personal pedagogical reasons. There are some references to a computer in Section 5.2 where Monte Carlo simulation is discussed and, to a lesser extent, in Section 8.2 where numerical solutions to differential equations are discussed.

Mathematics and physical science each had important effects on the development of the other. Mathematics is starting to play a greater role in the development of the life and social sciences, and these sciences are starting to influence the development of mathematics. This sort of interaction is extremely important if the proper mathematical tools are going to be developed for the various sciences. S. Bochner (1966) discusses the hand-in-hand development of mathematics and physical science. Some people feel that there is something deeper going on than simply an interaction leading to the formulation of appropriate mathematical and physical concepts. E. P. Wigner (1960) discusses this.

1.2. PROPERTIES OF MODELS

We begin with a definition based on the previous discussion: *A mathematical model is an abstract, simplified, mathematical construct related to a part of reality and created for a particular purpose.* Since a dozen different people are likely to come up with a dozen different definitions, don't take this one too seriously; rather, think of it as a crude starting point around which to build your own understanding of mathematical modeling.

We now have a problem: To fully appreciate the general discussion in the next two sections you should look at some concrete examples like those in Sections 1.4 and 1.5; however, you will need some abstract background to appreciate the examples fully. I suggest reading the remainder of the chapter through quickly and then coming back to this point and re-reading more carefully.

As far as a model is concerned, the world can be divided into three parts:

1. Things whose effects are neglected.
2. Things that affect the model but whose behavior the model is not designed to study.
3. Things the model is designed to study the behavior of.

The model completely ignores item 1. The constants, functions, and so on, that appear in item 2 are external and are referred to as *exogenous variables* (also called parameters, input, or independent variables). The things the model seeks to explain are *endogenous variables* (also called output or dependent variables). The exogenous-endogenous terminology is used in some areas of modeling. The input-output terminology is used in areas of modeling where the model is viewed as a box into which we feed information and obtain information from. The parameter-independent-dependent terminology is the standard mathematical usage.

Suppose we are hired by a firm to determine what the level of production should be to maximize profits. We would construct a model that enables us to express profits (the dependent variable) in terms of the level of production, the market situation, and whatever else we think is relevant (the independent variables). Next we would measure all the independent variables except the level of production and use the model to determine which value of the level of production gives the greatest profit.

Now let's look at things from the point of view of an economist who is seeking to explain the amount of goods firms produce. A two-part model could be constructed: Firms seek to maximize profits, and profits can be determined as sketched in the previous paragraph. In this model profits become an internal variable (of no interest except for the machinations of the model), and level of production changes from an independent to a dependent variable.

These three categories (neglected, input, and output) are important in modeling. If the wrong things are neglected, the model will be no good. If too much is taken into consideration, the resulting model will be hopelessly complex and probably require incredible amounts of data. Sometimes, in desperation a modeler neglects things not because he thinks they are unimportant, but because he cannot handle them and hopes that neglecting them will not invalidate the conclusions. A. Jensen (1966) discusses the development of a model for safety-at-sea problems. The main difficulty in formulating the model was to determine what types of encounters between ships were dangerous, that is, to separate items 1 and 2. He found this to be hard even with the aid of nautical experts. (If you want to know the answer, you'll have to read the article.)

Proper choice of dependent variables (i.e., output) is essential; we must seek to explain the things we can explain. Often this choice is relatively clear, as in the example involving the economist who wished to explain the level of production of a firm. Sometimes we need to be careful; for example, we could explain profits in terms of level of production, but not conversely as we might naively try to do, since we were asked to determine the best level of production.

Since different models make different types of simplifying assumptions, there is usually no single best model for describing a situation. R. Levins

(1968, p. 7) observed that "it is not possible to maximize simultaneously generality, realism, and precision." In the social sciences one is often content with a statement that something will increase; precision has been sacrificed for realism and (hopefully) generality. Simulation models usually try for precision and realism but sacrifice generality. These three trade-offs should become clearer after you have studied some actual models.

Definitions of the variables and their interrelations constitute the *assumptions* of the model. We then use the model to *draw conclusions* (i.e., to make predictions). This is a deductive process: *If the assumptions are true, the conclusions must also be true.* Hence a false prediction implies that the model is wrong in some respect. Unfortunately things are usually not this clear-cut. We know our model is only an approximation, so we cannot expect perfect predictions. How can we judge a model in this case?

A conclusion derived from a crude model is not very believable, especially if other models make contrary predictions. A result is *robust* if it can be derived from a variety of different models of the same situation, or from a rather general model. A prediction that depends on very special assumptions for its validity is *fragile*. The cruder the model, the less believable its fragile predictions.

You may notice that we have talked about conclusions, not *explanations*. Can a model provide explanations? This is a somewhat philosophical question, and different people have different notions of what constitutes an explanation. Let us grant that, in some sense, models can provide explanations. A decision about the validity of a model is usually based on the accuracy of its predictions. Unfortunately, two different models may make the same predictions but offer different explanations. How can this be?

We can think of the situation we are modeling as being a "black box" which outputs something for every input. ("Something" can be no output.) A model makes correct predictions if it outputs the model equivalent of the black box output whenever the model equivalent of the black box input is fed in. The mechanism is irrelevant when dealing with predictions, but the nature of the mechanism is the heart of an explanation. Although there is usually a situation in which two different models lead to different predictions, we may not be able to determine which prediction is correct. For example, a model of a politician can be constructed by assuming that his behavior is (1) motivated by concern for his fellow man or (2) motivated by a desire for public office. In many situations these two models lead to identical or very similar predictions. It may be difficult to make contradictory predictions that *can be checked*. Another example for those familiar with simple circuits is the mathematical equivalence between perfect springs and perfect LC circuits. Although the underlying mathematics is identical, no one would seriously suggest that Hooke's law for springs "explains" the circuit's behavior.

We have been talking about an ideal modeler. When any of us approaches a problem, we do so in a limited, biased fashion. The more open-minded, communicative, and creative we can be, the better our model is likely to be. The following poem illustrates the problems that can arise.

The Blind Men and the Elephant

It was six men of Indostan
    To learning much inclined,
Who went to see the Elephant
    (Though all of them were blind),
That each by observation
    Might satisfy his mind.

The First approached the Elephant,
    And happening to fall
Against his broad and sturdy side,
    At once began to bawl:
"God bless! but the Elephant
    Is very like a wall!"

The Second, feeling of the tusk,
    Cried, "Ho! what have we here
So very round and smooth and sharp?
    To me 'tis mighty clear
This wonder of an Elephant
    Is very like a spear!"

The third approached the animal,
    And happening to take
The squirming trunk within his hands,
    Thus boldly up and spake:
"I see," quoth he, "the Elephant
    Is very like a Snake!"

The Fourth reached out an eager hand,
    And felt about the knee.
"What most this wondrous beast is like
    Is mighty plain," quoth he;
"'Tis clear enough the Elephant
    Is very like a tree!"

The Fifth who chanced to touch the ear,
　　Said: "E'en the blindest man
Can tell what this resembles most;
　　Deny the fact who can,
This marvel of an Elephant
　　Is very like a fan!"

The Sixth no sooner had begun
　　About the beast to grope,
Than, seizing on the swinging tail
　　That fell within his scope,
"I see," quoth he, "the Elephant
　　Is very like a rope!"

And so these men of Indostan
　　Disputed loud and long,
Each in his own opinion
　　Exceeding stiff and strong.
Though each was partly in the right
　　And all were in the wrong!

　　　　　　John Godfrey Saxe (1816–1887)
　　　　　　Reprinted in Engineering Concepts
　　　　　　Curriculum Project (1971)

1.3. BUILDING A MODEL

Model building involves imagination and skill. Giving rules for doing it is like listing rules for being an artist; at best this provides a framework around which to build skills and develop imagination. It may be impossible to teach imagination. It won't try, but I hope this book provides an opportunity for your skills and imagination to grow. With these warnings, I present an outline of the modeling process.

1. **Formulate the Problem.**   What is it that you wish to know? The nature of the model you choose depends very much on what you want it to do.
2. **Outline the Model.**   At this stage you must separate the various parts of the universe into unimportant, exogenous, and endogenous. The interrelations among the variables must also be specified.

3. **Is It Useful?** Now stand back and look at what you have. Can you obtain the needed data and then use it in the model to make the predictions you want? If the answer is no, then you must reformulate the model (step 2) and perhaps even the problem (step 1). Note that "useful" does not mean reasonable or accurate; they come in step 4. It means: *If* the model fits the situation, will we be able to use it?

4. **Test the Model.** Use the model to make predictions that can be checked against data or common sense. It is not advisable to rely entirely on common sense, because it may well be wrong. *Start out with easy predictions*—don't waste time on involved calculations with a model that may be no good. If these predictions are bad and there are no mathematical errors, return to step 2 or step 1. If these predictions are acceptable, they should give you some feeling for the accuracy and range of applicability of the model. If they are less accurate than you anticipated, it is a good idea to try to understand why, since this may uncover implicit or false assumptions.

At this point the model is ready to be used. Don't go too far; it is dangerous to apply the model blindly to problems that differ greatly from those on which it was tested. Every application should be viewed as a test of the model.

You may not be able to carry out step 2 immediately, because it is not clear what factors can be neglected. Furthermore, it may not be clear how accurately the exogenous variables need to be determined. A common practice is to begin with a crude model and rough data estimates in order to see which factors need to be considered in the model and how accurately the exogenous variables must be determined.

Some models may require no data. If a model makes the same prediction regardless of the data, we are not getting something for nothing because this prediction is based on the assumptions of the model. To some extent, the distinction between data and assumptions is artificial. In an extreme case, a model may be so specialized that its data are all built into the assumptions.

Sometimes step 4 may be practically impossible to carry out. For example, how can we test a model of nuclear war? What do we do if we have two models of a nuclear war and they make different predictions? This can easily happen in fields of study that lack the precisely formulated laws found in the physical sciences. At this point experience is essential—not experience in mathematics but experience in the field being modeled. Even if predictions can be tested, the testing may be expensive to carry out and may require training in a particular field of experimental science. Since the absence of experimental verification leaves the modeling process incomplete, I have given test results whenever I have been able to obtain them.

1.4. AN EXAMPLE

We discuss models for the long term growth of a population in order to illustrate some of the ideas of the two previous sections. We want to predict how a population will grow numerically over a few generations. This is the problem (step 1 in Section 1.3).

Let the exogenous (independent) variables be the net reproduction rate r per individual, the time t, and the size of the population at $t = 0$. The net reproduction rate is the birth rate minus the death rate. In other words, it is the fractional rate of change of the population size: $r = (dN/dt)/N$. There is only one endogenous (dependent) variable, the size of the population at time t, which we denote by $N(t)$. We also refer to r as the *net growth rate*.

To obtain a simple model, we ignore time lag effects; that is, we assume that only the present value of N and its derivatives are relevant in determining the future values of N. (This will lead to a differential equation.) If the fraction of the population that is of reproductive age varies with t, this can be a very poor assumption. Let's also assume that the net reproduction rate r is a constant. This gives us a rather crude model with the basic relationship

$$(1) \qquad \frac{1}{N}\frac{dN}{dt} = r.$$

The model would certainly be useful if it fits the real world (step 3). The solution of (1) is $N(t) = N(0)e^{rt}$. Unless $r = 0$, the population will eventually either die out (r negative) or grow to fill the universe (r positive). Reasonable behavior of the population size is a very fragile prediction of the model. This casts serious doubt on the validity of using a constant net reproduction rate for predicting long term growth. This approach to a model illustrates an important point: *Study the behavior of your model in limiting cases* (in this case as time gets very long, i.e., as $t \to \infty$).

Our test of the model (step 4) for long term growth indicates that it must be rejected; however, it may be useful for short term predictions. Unfortunately, we specifically asked for long term predictions.

Clearly the growth rate of a population will depend on the size of the population because of such effects as exhaustion of the food supply. If the population becomes very large, we can expect the death rate to exceed the birth rate. Let's translate this into mathematics. We replace the net reproduction rate r in (1) by $r(N)$ which is a strictly decreasing function of N for large N and becomes negative when N is very large. Thus

$$(2) \qquad \frac{1}{N}\frac{dN}{dt} = r(N).$$

We've now redone step 2. The model is less useful than the previous one, because obtaining the exact form of $r(N)$ will be hard, perhaps even impossible. However, rough estimates can be obtained, so let's see what can be done with them. On to step 4. It can be shown that $N(t)$ approaches N_0, the solution of $r(N_0) = 0$, as time passes. This is a robust prediction, since we made very few assumptions concerning the nature of the function $r(N)$. Because the model was constructed to predict an upper limit for the size of a population, it is not surprising that it does so.

The cycle of steps 4, 2, and 3 can be repeated, since the model described by (2) has many drawbacks. For one thing, the population can only move closer to N_0 in the future. A real population often overshoots the steady state size N_0, and even steady state populations fluctuate slightly in size because of the somewhat random nature of births and deaths. One way to eliminate the first objection is to introduce *time lags*. For example, if the death rate m is not age dependent and the birth rate b changes from zero to a constant at age p, we could replace (1) by

$$(3) \qquad \frac{dN}{dt} = -mN(t) + bN(t - p).$$

The parameter p is called a time lag. Of course, we could make m ,and b functions of $N(t)$, $N(t - p)$, or some weighted average of N on the interval $[t - p, t]$. To allow for random fluctuations we must replace our deterministic model by a random one.

Another drawback is the assumption that it makes sense to talk about $r(N)$. If the age or sex ratios in a population are changing, this may be nonsense. To overcome this objection it is necessary to split the population into subpopulations based on age and sex. Demographic models are designed in this way: In a typical model time is broken up into discrete units such as 5 year periods, men are ignored, and women are divided into age classes separated by a single time unit. For each age class there is no longer simply a net birth rate but a death rate m_i and a birth rate b_i for female children. The number of newborn girls at time $t + 1$ is $N_0(t + 1) = \sum b_i N_i(t)$, and the number of women in class $i + 1$ is $N_{i+1}(t + 1) = (1 - m_i)N_i(t)$, the number surviving from class i at time t. Linear algebra is a natural tool for handling this model. Demographers frequently assume that b_i and m_i are independent of N, because they are interested in relatively short term predictions.

Seasonality may be important for short term models, since in many species births occur during a particular season and death rates are also dependent on the time of year. An explicit time dependence must be built into $r(N)$ to allow for seasonal effects. In a long term model encompassing many years we could probably avoid this complication by averaging birth and death rates over an entire year.

I hope this discussion makes it clear that we can't formulate an adequate model unless we know (1) what we hope to obtain from the model and (2) how complicated a model we are willing to tolerate. The latter is practically the same as how much data we are willing to supply, since complexity and data demands usually grow simultaneously.

1.5. ANOTHER EXAMPLE

The manager of a large commercial printing company asks your advice on how many salespeople to employ. Qualitatively, more salespeople will increase sales overhead, while fewer salespeople may mean losing potential customers. Thus there should be some optimum number. By "salespeople" I don't mean clerks, but people who travel, selling a company's products to other businesses; however, these ideas could be applied to salesclerks, too. This problem has been adapted from A. A. Brown et al. (1956). The original paper goes into greater depth than the following discussion and is well worth reading.

The problem as stated is unanswerable. What are the production limitations of the company? What are the goals of management? Maximum profit? Maximum "empire" with satisfactory profit? Something else? Unless these and similar questions are very clearly answered, recommendations may be quite inaccurate. A better approach would be to provide a description of the consequences of sales forces of various sizes. This would leave the final decision up to management, which is as it should be. To determine what effect a sales force will have, we must know what salespeople accomplish. Thus we can try to determine how salespeople spend their time and what results they obtain as a consequence of spending their time in that way. As long as salespeople need to be studied, we may as well ask: What is the best way (in terms of obtaining sales) for them to spend their time? We can then advise management on (1) how to obtain the greatest return from their sales force, and (2) the impact various sizes of sales forces will have on sales. This tentatively completes step 1.

Notice that we have changed the original problem considerably. We were asked, "How many salespeople should be employed?" Instead, we are going to answer two other questions which we formulated at the end of the previous paragraph. Actually the questions need further refinement. For example, different salespeople have different abilities, and their territories are probably different. The question how salespeople should spend their time contains a trap, because it invites us to ignore these variations. Again, if we change the size of the sales force, we can change the total geographical area covered, the effort expended per customer, or both. Thus the question on the consequences of various sizes of sales forces also contains potential traps. Clearly step 1 hasn't been completed; however, the best idea is probably to proceed and to realize that in studying a real situation we will eventually need to return to step 1 and formulate the questions more precisely in a way that depends greatly on the particular printing company being studied.

The major factor that will affect how much time a salesperson spends on a customer is what the salesperson can hope to gain. Observations indicate that businesses normally place most of their printing orders with one company. Hence we can classify customers as "in hand" or "potential." The former need to be held, and the latter need to be converted. In addition, we can classify customers according to how much money they have to spend. As an approximation we can assume (but it should be checked) that holding and conversion probabilities are independent of size. By running an experiment with the salespeople, or possibly by examining records if we are lucky, we can obtain an idea of how conversion and holding probabilities vary with the amount of time per week devoted to a customer. From this we can decide how a salesperson should spend their time, because one additional hour per month should produce the same expected gain in revenue regardless of which customer it is spent with. (If you don't see this, don't worry, I've omitted some details. Try rereading it after Chapter 4.) This completes steps 2 and 3 for the first part of the problem. We don't have the data to carry out step 4, but it should be relatively straightforward.

The decision on how a salesperson should divide his time together with the data on holding and conversion probabilities and data on the sizes of orders various businesses place will determine gross revenue as a function of number of salespeople. (Think about why this is so.)

The above outline indicates how we can attack the problem posed by management—remember: How many salespeople should we employ? The answer will consist of

1. A statement of how best to divide up a salesperson's time as a function of the number and type of customers being dealt with.
2. A table of expected gross income as a function of number of salespeople, assuming that the sales districts are divided up evenly.

The model building will not be complete until we actually collect the data and make predictions. As soon as we do this, we'll find that the data permit only rough estimates for items 1 and 2. Thus we should give some estimate of the *range* of the numbers: If we have n salespeople, the gross sales will be expected to be between X and Y dollars. We could also try to anticipate a question management is likely to raise:

> We can't make salespeople divide their time just the way you recommend. Besides, salespeople and customers are individuals. How sensitive are your recommendations to all this?

This example illustrates the importance of formulating the problem. The problem *as given* was hard or impossible to solve. By breaking it down and changing the goal (a tabulation of number of salespeople versus expected sales rather than simply an optimal number of salespeople), it became more approachable.

C. C. Lin and L. A. Segel (1974, Ch. 1) discuss applied mathematics and present two further examples. You may enjoy reading their chapter to obtain a somewhat different viewpoint. The first two chapters of J. Crank (1962) are also interesting reading. Chapters 2 and 3 of C. A. Lave and J. G. March (1975) present an interesting discussion of modeling.

PROBLEMS

Some of the problems in this book lead you step by step through the development of a model and thus resemble the mathematics problems you have seen in other courses; however, many problems are closer to real life: They are vaguely stated, have multiple answers (models), or are open ended. I strongly recommend working in small groups on the problems to bring out various ideas and evaluate them critically.

1. Suppose people enter the elevators in a skyscraper at random during the morning rush. The result will be several elevators stopping on each floor to discharge one or two passengers each.

    (a) Discuss schemes for improving the situation.
    (b) How could improvement be measured?
    (c) How could you model the situation to decide what scheme to adopt?

2. In the text we discussed models for the growth of a single population. Discuss models for the growth of two interacting populations. This problem has been phrased very vaguely, and before working on it at home decide on a more concrete situation (or situations) in class.

3. How far can a migrating bird fly without food?

4. If all five employees can run all six machines in your shop, how should you decide whom to assign to which job?

5. Discuss the differences and similarities in models of urban vehicular traffic that you would construct to deal with the following problems. To what extent could one model be used to handle problems it wasn't

designed for? Consider each case separately. Don't try to set up detailed models, just discuss your general approach.

(a) You are working for a citizens' committee which wants to convince the city council to ban private vehicles in the city because of pollution.

(b) The city council has asked you as a traffic engineering expert to study the possibilities of speeding up traffic flow by changing traffic signal times, setting up one-way streets, and anything else you can think of that will help the traffic problem, not upset the voters, and not cost much to implement.

(c) Since your recent efforts have won you a reputation, the city council has given you a contract to study the feasibility of banning private vehicles and taxis in most of the city as a means of reducing atmospheric and noise pollution, but in a fashion that won't interfere greatly with the mobility of the populace. Since this is a thorny problem with many conflicting goals—a political hornets' nest, the city fathers have told you to give them a straightforward recommendation so they can avoid the onus of decision making.

6. Unless you have been extremely lucky, you have had a large class in a poorly designed lecture hall.

(a) What are some criteria to be considered in designing a large lecture hall?

(b) One criterion is legibility of material written on the boards. Construct a model of legibility as a function of the distance your seat is from the board and the angle at which you look at the board. What will the curves of constant legibility look like on a floor plan? How can you test this prediction? Try it. Does this suggest shaping the back of the hall differently than is usually done? How?

(c) Can mathematical modeling help with any other criteria besides the one mentioned in (b)? Try to pick a criterion from among these possibilities and develop a model for it.

You may wish to look at A. A. Bartlett (1973) after working on this problem.

7. A common technique when no models are available is to collect data, try to fit curves, and then treat the curves as if they were a model or even an explanation. Discuss: Would you have faith in predictions made from such models? Explain. Two commonly misused techniques are *factor analysis* and *linear regression*. For a delightful spoof of the former, see J. S. Armstrong (1967).

8. One of the simplest models of population growth is the logistic equation $dN/dt = rN(1 - N/K)$.

   (a) Interpret r and K. Discuss the model.
   (b) Suppose you were given census data for a population (i.e., a table of date versus population size). How could you test the fit of the logistic model to the data? Remember that r and K are *not* given.
   (c) E. G. Leigh (1971, p. 124) quotes the following data from the U.S. Census Bureau on the growth of the U.S. population and from Gause on the growth of a population of the one-celled animal *Paramecium aurelia*. How well does the logistic model fit the data?

Year	$N \times 10^{-6}$	Day	N
1790	3.93	1	2
1810	7.24	2	7
1830	12.87	3	25
1850	23.19	4	68
1870	39.82	5	168
1890	62.95	6	138
1910	91.97	7	190
1930	122.78	10	122
1950	150.70	11	280
1970	208.0	12	260
		13	300

   (d) Can you suggest better models for the growth of the two populations given above? "Better" is a vague word. It could mean simpler, fitting the data more accurately, having a firmer biological and sociological foundation, and so on.

1.6. WHY STUDY MODELING?

Why not always deal with the real world instead of studying models? Modeling can avoid or reduce the need for costly, undesirable, or impossible experiments with the real world, as the following problems illustrate:

1. What is the most efficient way to divide the fuel between the stages of a multistage rocket?
2. What would be the effect of a very bad nuclear reactor accident?
3. How large a meteor was needed to produce Meteor Crater in Arizona?

In trying to "explain" the world, modeling is essential. Scientific theories are models and are frequently mathematical models. Every scientist from the purest to the most applied must know how to use such models whether he calls them that or not. For anyone planning to use mathematical models, an understanding of how to go back and forth between the world we live in and the world of mathematics is essential. This is the crux of mathematical modeling and this is what I hope this course will help you learn to do. It is neither science nor mathematics, but rather how to put them together. Science and mathematics courses are essential (you need something to put together), and this is no substitute for them.

PART 1
ELEMENTARY METHODS

CHAPTER 2

ARGUMENTS FROM SCALE

In this chapter we consider arguments based on proportionality. For example, if you make a scale model of an object with a scale of $1:l$, surface area will have a scale of $1:l^2$ and the volume a scale of $1:l^3$. Models using this sort of idea are discussed in the first section. The second section is based on the observation that physical laws remain the same if the units of measurement are changed.

2.1. EFFECTS OF SIZE

Cost of Packaging

Consider a product like flour, detergent, or jam, which is packaged in containers of various sizes. You've probably noticed that larger packages of such products usually cost less per ounce. This is often attributed to savings in the cost of packaging and handling. Is this in fact the major cause or are there other important factors? We try to see where this idea leads by constructing a simple model.

The cost of a product is the endogenous variable. We are interested in seeing how it varies with the exogenous variable, size. Cost clearly depends on competition and the scale of the business. We neglect these factors and concentrate on expenses due to materials and handling. Since we are neglecting some important factors (name some), the resulting predictions will be crude. In addition, there are various constants involved which we do not even pretend to evaluate.

Let's begin by studying the wholesale cost, that is, the price the retailer pays for the product. This is a sum of several costs plus various profit markups by middlemen. Since profit markups are usually in terms of percentages, we can absorb them in constants later; for example, a 30 % markup multiplies constants by 1.30. The main costs that enter the wholesale price are:

1. Cost of producing the product, a.
2. Cost of packaging the product, b.
3. Cost of shipping the product, c.
4. Cost of the packaging material, d.

We will consider each of these in turn.

It is reasonable to assume that a is proportional to the amount of the good being produced. We write this as $a \propto w$, which is read "a is proportional to the weight w."

The costs of packaging depend on how long it takes to fill the package, how long it takes to close the package, and how long it takes to load the package into a box for shipping. The first time is probably nearly proportional to the volume (hence the weight), while the latter two times are probably about the same for all sizes of packages in a reasonable range. Thus $b \approx fw + g$ for some positive constants f and g. (The symbol $\approx$ means "approximately equal to.")

Shipping charges may depend on both weight and volume. Since volume is proportional to weight for filled packages, we have $c \propto w$.

The cost of the packaging material is more complicated. It depends on the costs the package manufacturer must meet. Thus we must consider a, b, c, and d for the package manufacturer. We neglect d; that is, we neglect the cost of the containers for the material from which the final packages are made. From the analysis we have just completed, the cost per package depends on the weight and volume of the package. If the range of packages we are considering is not too large, it is reasonable to assume that the packaging material is the same for all sizes of packages. Therefore the amount of material per package (hence the weight of a package) is proportional to the area of the surface to be covered. The volume per package is proportional to either the surface area or the volume of the package, depending on whether the packaging is shipped collapsed (like cardboard) or preformed (like glass). Therefore the expenses per package of the package supplier are $hw + kS + m$, for constants $h \geq 0$, $k > 0$, and $m > 0$, where S is the surface area. Except for a markup, this is the cost d to the packager.

We now use a scale argument to reduce everything to one independent variable, weight. Let us assume that the various packages are roughly geometrically similar. The volume is nearly proportional to the cube of a linear dimension, and the surface area is nearly proportional to the square of a linear dimension: $v \propto l^3$ and $S \propto l^2$. Hence $S \propto v^{2/3}$. Since $v \propto w$, we have $S \propto w^{2/3}$. Thus the *wholesale* cost per ounce is

(1) $$\frac{\text{Cost}}{w} = \frac{a + b + c + d}{w} = n + pw^{-1/3} + \frac{q}{w},$$

for positive constants n, p, and q. From this we see that the cost per ounce decreases as the size of the package increases, in agreement with the observation made at the start of this discussion.

Can we make any interesting predictions? Given three different costs and weights, we could solve for n, p, and q in (1) and use the results to predict the prices for packages of other sizes. Because of the crudity of our model, it is unlikely that our equation will fit very well. We should not take the exact form of (1) too seriously. Another way to fit a curve, which allows for inaccuracies, is the method of least squares. For this to be a reasonable test of the model, we should have more data points than parameters. Since (1) involves three constants, we should have more than three values for the cost and weight of a single product. This is hard to obtain because of the limited number of different-sized packages in which a particular product is available. Therefore we need a different approach.

The cost per ounce *decreases* at a rate

(2) $$r = -\frac{d(\text{cost}/w)}{dw} = \frac{p}{3w^{4/3}} + \frac{q}{w^2}.$$

This is a decreasing function of w. Thus the increase in the rate of savings per ounce is less when the package is larger. We can also compute the rate of total savings:

$$rw = \frac{pw^{-1/3}}{3} + qw^{-1}.$$

It is also a decreasing function of w.

The consumer is not likely to understand this. We can make a statement like (2) in simpler terms:

> In purchasing prepackaged products, doubling the size of the package purchased tends to result in greater savings per ounce when the packages are small than when they are large.

You can prove this by taking the difference of (1) at w and $2w$ and verifying that it is a decreasing function of w. We have said "tends to" because the model is crude.

These predictions seem to rely heavily on the exact form of (1). Actually qualitative predictions like these are usually quite robust. It would be desirable to derive them from a more general model if we wished to pursue the model more seriously, but I don't know how to do this and I don't think that the problem is worth the effort.

This discussion concerned wholesale prices. What about retail prices? The retailer's costs depend on wholesale prices and handling and storage costs. As above, the latter two costs are of the form $Hw + M$. If the wholesaler sets his price at a fixed percentage above his costs, then we again obtain an equation of the form (1). The conclusions we reached above are therefore valid for retail prices too.

In Problem 1 you are asked to study the model further and test it against actual data.

Speed of Racing Shells

In the college sport of crew racing the best times vary from class to class. Why? Can we advise a coach how to adjust the shells so that he can pit his teams against each other on an equal basis in practice? This model is adapted from T. A. McMahon's article (1971) and deals with data for men only.

Racing shells are boats propelled by oarsmen in sporting contests. They hold one, two, four, or eight oarsmen and are built to certain specifications. Figure 1 is a rough diagram of a racing shell. For an eight-man crew there is a lightweight category and a heavyweight category. Heavyweight

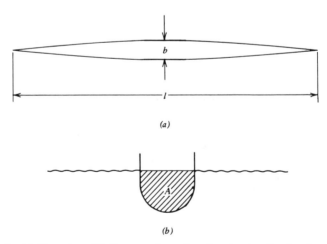

(a)

(b)

Figure 1   (a) Top view. (b) Cross section of center. l = length; b = beam; A = cross-sectional area.

Table 1   Times of Racing Crews in Four Meets

Number of men	I	II	III	IV
8	5.87	5.92	5.82	5.73
4	6.33	6.42	6.48	6.13
2	6.87	6.92	6.95	6.77
1	7.16	7.25	7.28	7.17

oarsmen average about 86 kilograms, and lightweight oarsmen about 73 kilograms. This gives five classes. (There are others which we ignore because of a lack of data.) McMahon observed that there is a rather consistent difference between the best times of the various classes. Table 1 lists the information he presented on best times for 2000 meter races in four international competitions. The eight-man entry is the heavyweight time. McMahon also states that the time of an eight-man heavyweight crew is about 5% better than the time of an eight-man lightweight crew.

We want to explain all this.

Rather than present the underlying assumptions of the model in one ad hoc package, we develop them as we proceed.

A shell is propelled by the power of the oarsmen and retarded by the drag of the water. The balance of these two forces determines the speed of the shell, hence its time in the race. We assume

1. The only drag force on the shell is due to skin friction and this force is proportional to Sv^2, where S is the wetted surface area and v is the velocity.

The expression for the skin friction drag given in the assumption is obtained from hydrodynamics. The power P required to maintain velocity v is, by definition, equal to the drag force times the velocity. Hence $P \propto Sv^3$, and so $v \propto (P/S)^{1/3}$

We assume

2. The oarsmen in the shell all have the same weight and the same constant power output for the entire course of the race.

It follows that v is constant, except for the brief period when the shells are starting up. Hence the course time t is proportional to v^{-1}, and so

(3)
$$t \propto \left(\frac{S}{P}\right)^{1/3}$$

We now consider the time difference between the heavyweight and lightweight eight-man crews. We want to explain it and then see if we can find a way to redesign the shells so that the two classes will be more nearly equal.

The subscripts H and L denote heavyweight and lightweight, respectively. From (3) we obtain

$$(4) \qquad \frac{t_L}{t_H} = \left(\frac{S_L}{S_H}\right)^{1/3} \left(\frac{P_H}{P_L}\right)^{1/3}.$$

We must say something about power output and wetted surface area if we are going to explain the 5% edge of the heavyweight team. Unfortunately power output information is not obtainable; however, we know that the ratio of the weights of heavyweight and lightweight oarsmen is about 86 kilograms/73 kilograms $= 1.18$. Therefore we try to relate power and weight.

Sustained power output depends on such factors as lung volume (actually lung surface area, but this is proportional to volume because the lungs consist of small cells whose size is independent of the size of the person) and muscle volume. For *similarly proportioned people*, these are proportional to the total weight. Hence we can expect power output to be proportional to the weight w of an oarsman times the number of oarsmen. Since $w_H/w_L = 1.18$ and both shells have eight oarsmen, $P_H/P_L = 1.18$. Combining this with (4),

$$(5) \qquad \frac{t_L}{t_H} = (1.18)^{1/3}\left(\frac{S_L}{S_H}\right)^{1/3} = 1.06\left(\frac{S_L}{S_H}\right)^{1/3}.$$

If we make the rough assumption that $S_L = S_H$, then (5) comes close to the 5% observed difference. Actually the surface area for a loaded heavyweight shell is slightly greater than that for a lightweight shell. When this is taken into account, the 6% edge in (5) decreases slightly. We haven't predicted the edge precisely, but we have explained why it is in the neighborhood of 5%.

How can the shells be redesigned to achieve equality? For fixed power output we obtain $t \propto S^{1/3}$ from (3). To change the time we must change the wetted surface area of the loaded shell. Let the subscripts p and r denote the present and redesigned shells, respectively. Then

$$\frac{S_r}{S_p} = \left(\frac{t_r}{t_p}\right)^3.$$

The lightweight crews will have times about equal to those of the heavyweight crews if $t_r/t_p = 0.95$. By the above equation, $S_r/S_p = 0.86$. In words, the wetted surface area of a loaded lightweight shell should be decreased by about 14%, or we could slow the heavyweights down by an increase of wetted area of about 16% $(1/0.86 = 1.16)$.

We now compare the times of various-sized shells by expressing the endogenous variable, course time, in terms of the exogenous variable, team size. To do this we have to relate S and P to the size of the team. If assumption 2 is extended to all oarsmen in all shells, the power will be proportional to the number of oarsmen n. Hence (3) reduces to

$$(6) \qquad\qquad t \propto \left(\frac{S}{n}\right)^{1/3}.$$

We need some information about the relative sizes of the various shells so that we can compute S. The information in Table 2 was presented by McMahon as evidence for the assumption:

3. The shells are geometrically similar, and their loaded weights are proportional to n. Furthermore, the submerged parts of the loaded shells are also geometrically similar.

Table 2   Shell Design Parameters

n	l	b	l/b	weight/n
8	18.28	0.610	30.0	14.7
4	11.75	0.574	21.0	18.1
2	9.76	0.356	27.4	13.6
1	7.93	0.293	27.0	16.3

Note: l = length; b = beam.

The variation in the "l/b" and "weight/n" columns shows that this is a rather crude assumption, but it is about the best we can do, since a table of wetted surface areas is not available.

The volume of water displaced by a shell is proportional to its total weight. This volume is also proportional to lA. By assumption 3, weight is proportional to the number of oarsmen n, and $A \propto l^2$. Thus

$$(7) \qquad\qquad n \propto lA \propto l^3.$$

The values of l and n listed in Table 2 do not satisfy $n \propto l^3$. Therefore the similarity assumption is wrong. What can we do about it? For the sake of continuity, we postpone discussing this problem.

The total submerged surface area is proportional to l times the submerged perimeter of cross section A in Figure 1. By assumption 3, this perimeter is proportional to $A^{1/2}$ which is in turn proportional to l. Thus $S \propto l^2$. From (7) we obtain $S \propto n^{2/3}$, and so (6) becomes

(8) $$t \propto n^{-1/9}$$

This yields the prediction:

> Times are proportional to the number of oarsmen raised to the power $-\frac{1}{9}$.

We can test this prediction by graphing t versus n in some fashion. It is much easier to see if points are close to a straight line than it is to see if they are close to a curve. For this reason relationships like (8) are usually plotted on what is called log–log paper. It gives the effect of plotting $\log n$ against $\log t$, which equals $c - \log n/9$ if (8) is correct. If you do this, you will discover that the points come close to lying on a straight line of slope $-\frac{1}{9}$ as predicted. For a least squares curve fit, see Section A.7, especially page 237.

We are in an awkward situation: the prediction in (8) has been verified, but the intermediate result in (7) is wrong. One possible explanation for this is that the central portions of the shells (which displace most of the water, hence are the most important) obey the similarity assumptions better than the ends of the shells. I do not have the data to check this possibility. This central length λ and the cross section enter into the calculations for volume and surface area. A reasonable rough approximation is that volume and surface area are proportional to λ^3 and λ^2, respectively. The previous calculations can then be carried out with λ replacing l.

We can give a more robust argument that leads to (8). The volume of the submerged portion of the shell is proportional to the weight of the loaded shell by Archimedes' law. The weight is very nearly proportional to n. Hence the volume is very nearly proportional to n. Since the shells are all approximately the same shape, the surface area is nearly proportional to the $\frac{2}{3}$ power of the volume. Hence S is nearly proportional to $n^{2/3}$. By (6), $t \propto (n^{2/3}/n)^{1/3} = n^{-1/9}$. The important point in this argument is that surface area tends to remain proportional to the $\frac{2}{3}$ power of the volume, even when the shape varies somewhat from shell to shell. Thus we do not need the exact similarity assumption 3.

Size Effects in Animals

Why do animals have the proportions they do? You may have noticed that larger animals tend to have stockier bodies and relatively heavier legs. For instance, a deer is not a scale model of an elephant even if we neglect superfluous things like the head and the pelt. Why is the largest bird much smaller than a large mammal? Why can fleas jump so high relative to their size? (Is this the basis of flea circuses?)

Various people have applied proportionality arguments to biology. The books by N. Rashevsky (1960, pp. 251–275) and J. Maynard Smith (1968, pp. 6–17) contain a variety of examples from which the following discussion was adapted. You may also wish to read J. B. S. Haldane (1928). K. Schmidt-Nielson's book (1972) is worth reading, but only a small part of it deals with scaling problems.

We want to study how the size of a quadruped affects its locomotion and the proportions of its body and limbs. The only locomotion question we consider is jumping. J. Maynard Smith (1968, p. 12) has observed that the height to which a jumping mammal can leap seems to be nearly independent of its size. In particular, he notes that a jerboa (a mouselike rodent) and a kangaroo can jump about equally high. We want to obtain some idea of what this may mean. If you wish a fuller exposition of movement, see the books mentioned above.

The structure of animals is quite complex, and so it is easy to build very involved models. Rather than becoming lost in a morass of complicated, uninterpretable results, we use very crude models. At a couple of critical points we'll unfortunately have to rely on some results from elasticity theory.

We now study how the dimensions of the body (trunk) of an animal are related to its weight. As a crude approximation, we think of the trunk of the animal as a flexible beam supported at the ends by the legs. Flexible beams have been well studied in elasticity theory, so there are results ready for our use. If a beam of length l, vertical thickness t, and cross-sectional area A is subjected to a uniform load F while its end points are held fixed, a result from elasticity theory states that the maximum deflection δ satisfies

$$\delta \propto \frac{Fl^3}{t^2 A}.$$

The force F is due mainly to the weight of the trunk, which is roughly proportional to lA. Using this we see that

$$(9) \qquad \frac{\delta}{l} \propto \frac{l^3}{t^2},$$

where δ/l is the relative sagging. It is reasonable to suppose that there exists some physically determined upper limit to δ/l above which the animal's trunk will be cripplingly deformed. Some dog breeds (e.g., St. Bernard) may be at this limit. When δ/l is much below this limit, body material is being used unnecessarily for support. It is reasonable to suppose that such an inefficient use of body material is eliminated by evolution. Hence we treat δ/l as a constant. From (9) we obtain

$$(10) \qquad t \propto l^{3/2};$$

that is, larger animals have relatively thicker trunks. Rashevsky (1960, vol. 2, p. 263) has plotted log t against log l and found fair agreement with (10). The mass m of the trunk is roughly proportional to lA. Since most animals have roughly similar cross sections, $A \propto t^2$. Thus $m \propto lt^2$ and so $m \propto l^4$ by (10). Combining these observations gives

$$(11) \qquad l \propto m^{1/4}, \qquad t \propto m^{3/8}, \qquad \frac{t}{l} \propto m^{1/8}.$$

Interpret these results.

How does limb size vary with body weight? Our model here is even cruder than the previous one. The leg bones must be strong enough to withstand the bending strain put on them when the animal moves. From elasticity theory, the ability of a bone to withstand a force is proportional to its cross sectional area A_b. Force equals mass times acceleration. For slow moving animals, acceleration is mostly due to gravity. For fast moving animals, accelerations are still about equal because they depend on the rate of muscle contraction, which has about the same maximum value in all species. Thus the force applied is proportional to the mass m of the animal, and so $A_b \propto m$. If d is the diameter of the leg bone, $d^2 \propto A_b$ and so $d \propto m^{1/2}$. Note that, if everything remained in proportion for animals of different sizes, we would have $d \propto m^{1/3}$. Hence our model predicts that bone diameter increases faster than proportionally; that is, the legs of larger animals are relatively thicker than the legs of smaller animals.

How does the height an animal can jump depend on its size? To jump a height h an animal of mass m must do an amount of work proportional to mh. This work is accomplished by the muscles as the legs move from a crouched position at the start to a stretched position just before leaving the ground. The work that can be done by a muscle is proportional to its volume V_m. Thus mh is proportional to V_m, and so

$$(12) \qquad h \propto \frac{V_m}{m}.$$

If we make the plausible assumption that V_m is proportional to m, the total mass of the animal, we obtain $h = $ constant from (12). However, it also seems plausible to assume that the cross sectional area of the muscle is proportional to A_b, the cross sectional area of the leg bones. Since $A_b \propto m$, it follows from (12) that λ, the length of the muscle, is proportional to h. Since λ increases with size, this leads to the conclusion that h increases with size. Which approach is wrong and why? Actually, neither is correct. Rather than make "plausible assumptions" in a naive fashion, we need to look at the situation structurally: What is it that determines the size of leg muscles? If the muscles are too strong, they will cause the leg bones or joints to break. A plausible but somewhat technical bioengineering argument leads to the conclusion that, if bone breakage is the major consideration, $V_m \propto A_b$. Thus (12) becomes

(13)
$$h \propto \frac{A_b}{m}.$$

If we accept our earlier conclusion that $A_b \propto m$, we obtain $h = $ constant. This conclusion was based on the idea that the importance of leg bone cross section derived from supporting the animal; however, we see from (13) that for jumping mammals the importance of leg bone cross section may derive from the height the animal wishes to jump.

It would be interesting to study a table of h, A_b, V_m, and m for jumping mammals. I have been unable to locate such data. In fact, not many data are available to test our size effect models. Of course, one can always measure photographs or actual animals. Perhaps you'd like to do it. Besides the graphical data given by Rashevsky mentioned earlier, T. A. McMahon (1973) presents further graphical data, and D. D. Davis (1962) notes that in domestic cats and lions structures associated with locomotion satisfy mass relationships of the form $w \propto m^r$, where $r \geq 1$, while structures associated with metabolism have $r < 1$. W. R. Stahl and J. Y. Gummerson (1967) analyzed five species of primates (tamarins, squirrel monkeys, vervet monkeys, macaques, and baboons). Among their results are the following 95 % confidence estimates for r in $x \propto m^r$.

x	r
Trunk height	0.26–0.29
Chest circumference	0.35–0.38
Thoracic width	0.27–0.35
Midshaft humerus diameter	0.39–0.45

It was not clear to me what "trunk height" meant. The first two measurements fit the l and t results in (11) quite well, but the thoracic width does not fit the $t \propto m^{3/8}$ prediction. The humerus diameter measurement leads to a value of r in $A_b \propto m^r$ considerably less than the predicted value of 1.

PROBLEMS

1. This problem relates to the model of the cost of packaging. The conclusion drawn from (1) that costs per ounce for larger packages are less holds for the data given at the end of this problem, but this is a relatively crude result. Equation (2) cannot be checked, because we cannot compute derivatives, only differences. Moreover, the rule on doubling the size of a package cannot usually be checked, since manufacturers tend to package

products in odd sizes. We want a more flexible form of the doubling rule, and so we shall derive a finite difference analog of (2).

(a) Let $w_1 < w_2 < w_3$ be the weights of various-sized packages of a packaged product and c_1, c_2, and c_3 the costs *per ounce* of the packaged product. Derive the following result.

$$\frac{c_1 - c_2}{w_2 - w_1} = \frac{q}{w_1 w_2} + \frac{p}{w_1^{1/3} w_2 + (w_1 w_2)^{2/3} + w_1 w_2^{1/3}} > \frac{c_2 - c_3}{w_3 - w_2}$$

Why is this analogous to the statement that r is a decreasing function of w?

(b) The following data was collected at random in a supermarket in 1972. Test the result given in (a). The samples in each group came from the same store at the same time and were of the same brand. The packages within a group appeared similar except for the 12 and 32 ounce ketchup bottles. The former was labeled "wide mouth" and the latter was labeled "jug." It may be relevant that the 5 and 10 pound bags of flour were on a shelf marked "new low price." The data in each table is taken from a single brand.

<table>
<tr><th colspan="2" align="center">Ketchup</th><th colspan="2" align="center">Powdered Milk</th></tr>
<tr><th>Ounces</th><th>$</th><th>Quarts</th><th>$</th></tr>
<tr><td>12</td><td>0.29</td><td>3</td><td>0.49</td></tr>
<tr><td>14</td><td>0.26</td><td>8</td><td>1.09</td></tr>
<tr><td>20</td><td>0.36</td><td>14</td><td>1.59</td></tr>
<tr><td>32</td><td>0.57</td><td>20</td><td>2.19</td></tr>
</table>

<table>
<tr><th colspan="2" align="center">Tomato Sauce</th><th colspan="2" align="center">Flour</th></tr>
<tr><th>Ounces</th><th>$</th><th>Pounds</th><th>$</th></tr>
<tr><td>8</td><td>0.15</td><td>2</td><td>0.27</td></tr>
<tr><td>15</td><td>0.25</td><td>5</td><td>0.39</td></tr>
<tr><td>29</td><td>0.45</td><td>10</td><td>0.85</td></tr>
</table>

<table>
<tr><th colspan="3" align="center">Detergent Powder</th></tr>
<tr><th>Pounds</th><th>Ounces</th><th>$</th></tr>
<tr><td>3</td><td>1</td><td>0.81</td></tr>
<tr><td>5</td><td>4</td><td>1.29</td></tr>
<tr><td>10</td><td>11</td><td>2.52</td></tr>
</table>

(c)   To test the model further it would be desirable to make additional predictions that could be checked against the data. Can you make a testable prediction analogous to the statement that rw is a decreasing function of w? Can you obtain any other qualitative predictions from the model, which can be tested with the data?

2.   Can you think of any data that it would be reasonable to try to obtain and which would allow you to improve the model of the speed of racing shells?

3.   T. A. McMahon has suggested that, if the lightweight eight-man shell were a scale model of the heavyweight eight-man shell when loaded [i.e., if the dimensions had the ratio $1:(1.18)^{1/3}$], the 5% edge would be eliminated. Do you agree with this? Why? (Recall that we needed a ratio of redesigned to present surface areas of 0.86.)

4.   Smaller mammals and birds have faster heart rates than larger ones. If we assume that evolution has determined the best rate for each, why isn't there one single best rate? Is there a model that leads to a correct rule relating heart rates? A warm-blooded animal uses large quantities of energy in order to maintain body temperature, because of heat loss through its body surfaces. Since cold-blooded animals require very little energy when they are resting, the major energy drain on a resting warm-blooded animal seems to be maintenance of body temperature. Let's explore a model based on this idea.

The amount of energy available is roughly proportional to blood flow through the lungs—the source of oxygen. Assuming the least amount of blood needed is circulated, the amount of available energy will equal the amount used.

(a)   Set up a model relating body weight to basal (resting) blood flow through the heart. Use the data below to check your model.
(b)   There are many animals for which pulse rate data is available but not blood flow data. Set up a model that relates body weight to basal pulse rate. What sort of assumptions do you need to make about hearts? How could they be checked? Use the data below to check your model.
(c)   Discuss the discrepencies that arise in testing your models in (a) and (b).

After working on the model you may wish to read M. Kleiber (1961, Ch. 10, especially pp. 199–209). It would be good if someone did this and reported on it.

Data on Mammals (Altman and Dittmer, 1964, pp. 234–235)

Mammal	Weight (kilograms)	Pulse (beats per minute)
Shrew	0.003–0.004	782
Bat	0.006	588
Mouse	0.017	500
Hamster	0.103	347
Kitten	0.117	300
Rat	0.252	352
Guinea pig	0.437	269
Rabbit	1.34	251
Opussum	2.2–3.2	187
Seal	20–25	100
Goat	33	81
Sheep	50	70–80
Swine	100	60–80
Horse	380–450	34–55
Cattle	500	46–53
Elephant	2,000–3,000	25–50

Note: Rates may not be basal.

Data on Humans (Spector, 1956, p. 279)

Age	5	10	16	25	33	47	60
Weight (kilograms)	18	31	66	68	70	72	70
Pulse (beats per minute)	96	90	60	65	68	72	80
Blood flow through heart (deciliters per minute)	23	33	52	51	43	40	46

Data on Some Mammals (Spector, 1956, p. 279)

	Rabbit	Goat	Dog	Dog	Dog
Weight (kilograms)	4.1	24	16	12	6.4
Blood flow through heart (deciliters per minute)	5.3	31	22	12	11

Data on Small Birds (Altman and Dittmer, 1964, p. 235)

Bird	Weight (grams)	Pulse (beats per minute)
Hummingbird	4	615
Wren	11	450
Canary	16	514
Sparrow	28	350
Dove	130	135

Data on Large Birds (Altman and Dittmer, 1964, p. 235)

Bird	Weight (grams)	Pulse (beats per minute)
Gull	388	401
Chicken	1,980	312
Vulture	8,310	199
Turkey	8,750	93
Ostrich	80,000	65

Note: Rates may not be basal.

5. In *Gulliver's Travels*, the Lilliputians decided to feed Gulliver 1728 times as much food as a Lilliputian ate. They reasoned that, since Gulliver was 12 times their height, his volume was $12^3 = 1728$ times the volume of a Lilliputian and so he required 1728 times the amount of food one of them ate. Why was their reasoning wrong? What is the correct answer?

6. When you hear something, how does the apparent intensity vary with the actual intensity? What about brightness, weight, and so on? In the nineteenth century Weber formulated a law stating that the just noticeable difference (jnd) in signal intensity is proportional to the intensity of the signal. The constant of proportionality k varies from 0.003 for pitch to 0.2 for salinity. Fechner took Weber's law and assumed that all jnd's were psychologically equal for a given type of stimulus. This led to the Weber–Fechner law relating psychological intensity S, measured in jnd's, to physical intensity F: $S = g(F)$.

   (a) Show that Weber's law states that, if S_1 and S_2 differ by 1 jnd, $\log F_1$ and $\log F_2$ differ by some constant k. Derive the Weber–Fechner law: If S_1 and S_2 differ by an integral number N of jnd's,

$\log F_1$ and $\log F_2$ differ by kN. Conclude that $g(F) = k \log F + C$. Sound loudness and star brightness are both measured in logarithms of energy (decibels and magnitude). Why is this done?

(b)   Conclude from the Weber–Fechner law that, if $F_1/F_2 = F_3/F_4$, then $S_1 - S_2 = S_3 - S_4$; that is, the apparent intensities as measured by a person seem to differ by the same amount. This result is usually fairly accurate for intermediate values of intensity but is often inaccurate at extremes. However, Weber's law is usually fairly accurate over the entire range. How can this be? (Find the hidden assumption in Fechner's derivation.)

(c)   Stevens discovered that equal ratios of physical intensity correspond to equal ratios of psychological intensity; that is,

$$\frac{F_1}{F_2} = \frac{F_3}{F_4} \quad \text{if and only if} \quad \frac{S_1}{S_2} = \frac{S_3}{S_4}.$$

Let $s_i = \log S_i$ and $f_i = \log F_i$. Suppose $f_2 = f_1 + \delta$ and $f_4 = f_3 + \delta$. Letting $\delta \to 0$, show that ds/df is a constant. Describe the function $S = g(F)$ in Stevens' law.

(d)   How can Weber's law and Stevens' law both be nearly true?

Various psychology texts discuss the subject of this problem, for example, E. Fantino and G. S. Reynolds (1975, pp. 220–226). For a more extended discussion see S. S. Stevens (1974, Ch. 1). A. Rapoport (1976) discusses this problem and other topics in mathematical psychology.

7. Atmospheric drag is roughly proportional to Sv^2, where S is surface area and v is speed, for many common objects (e.g., moving cars and falling bodies).

(a)   If v is the terminal velocity of a falling object, show that for similarly proportioned objects $v \propto m^{1/6}$.

(b)   Show that on collision with the ground the kinetic energy per unit area that must be converted into some other form of energy is proportional to $m^{2/3}$.

(c)   Discuss the effect of falling on animals of various sizes. Remember that larger animals have larger bones. (See page 28.)

2.2. DIMENSIONAL ANALYSIS

Dimensional analysis is a tool of the physical sciences. It is based on the observation that physical quantities have dimensions associated with them and that physical laws remain unaltered when the fundamental units for measuring the dimensions are changed. For example, the area of a rectangle is the base times the height regardless of whether we measure in feet or meters as long as the units of area are (feet)^2 or (meters)^2, respectively.

Dimensional analysis alone will not give the exact form of a function, but it can lead to a significant reduction in the number of variables. As a result, it may be much easier to prepare tables of a function experimentally. A related usage of dimensional analysis is the design of scale models: It helps you face the problem of how to scale the physical parameters of the system so that predictions can be made for the real problem by analyzing the behavior of the scale model.

The examples presented here are adapted from L. I. Sedov (1959). The first book on the subject was written by P. W. Bridgman (1931). J. F. Douglas (1969) gives a recent, standard, elementary introduction to the subject. If you would like to read a text containing problems with solutions, see H. L. Langhaar (1951). S. J. Kline (1965) presents a critical introduction to dimensional analysis and related topics.

Theoretical Background

The basic physical dimensions are usually mass, length and time. We denote them by M, L, and T. Since we can measure velocity in feet per second, it has the dimension of length/time. We express this by saying that the dimension of velocity is L/T. By Newton's law, force equals $d(mv)/dt$, where m is mass, v velocity, and t time. Hence it has the dimension of mv/t, which is $M(L/T)/T = MLT^{-2}$.

If all the terms in an equation have the same dimension, we say that the equation is *dimensionally homogeneous*. By our definition of the dimension of force, we have made Newton's law dimensionally homogeneous.

Consider Newton's law of gravitation:

(14) $$F = \frac{Gm_1 m_2}{r^2},$$

where G is a universal constant, m_1 and m_2 are the masses of two bodies, and r is the distance between them. We have just determined that the dimension of the left hand side is MLT^{-2}. The dimension of $m_1 m_2/r^2$ is $M^2 L^{-2}$. The two sides of the equation apparently have different dimensions. Actually, the value of the constant G depends on the units of measurement and so is also given a dimension. To make the law of gravitation dimensionally homogeneous, the dimension of G must be

$$\frac{MLT^{-2}}{M^2 L^{-2}} = M^{-1} L^3 T^{-2}.$$

By assigning dimensions to variables and constants in this way, we can make *all the laws of physics dimensionally homogeneous.* This is not as surprising as it may sound. Almost everyone is aware to some extent that it is not correct to compare things that have different dimensions.

The basic theorem of dimensional analysis is the Buckingham pi theorem. It can be stated as follows.

THEOREM.   An equation is dimensionally homogeneous if and only if it can be put in the form

$$f(\pi_1, \pi_2, \ldots) = 0,$$

where f is some function and $\pi_1, \pi_2, \ldots$ are dimensionless products (and quotients) of the variables and constants appearing in the original equation. Not all dimensionless products need to be included in the list $\pi_1, \pi_2, \ldots$. Only a set from which all others can be formed by multiplication and division is needed.

It can be shown that the number of products in the list $\pi_1, \pi_2, \ldots$ need not exceed the number of variables and physical constants in the original equation.

As an example of the theorem we return to the law of gravitation (1). Consider a product of the form

$$\pi = G^a m_1^b m_2^c r^d F^e,$$

where the exponents a, b, c, d, and e are arbitrary. The dimension of this product is

$$(M^{-1}L^3T^{-2})^a M^b M^c L^d (MLT^{-2})^e = M^{b+c+e-a} L^{3a+d+e} T^{-2(a+e)}$$

From this we see that π is dimensionless if and only if

$$b + c + e - a = 0, \qquad 3a + d + e = 0, \qquad a + e = 0.$$

We can choose a and b arbitrarily. Then $c = 2a - b, d = -2a$, and $e = -a$. Since $(a, b) = a(1, 0) + b(0, 1)$, all dimensionless products can be obtained from the two cases $(a, b) = (1, 0)$ and $(a, b) = (0, 1)$. These give

$$\pi_1 = \frac{Gm_2^2}{r^2 F}, \qquad \pi_2 = \frac{m_1}{m_2}.$$

Buckingham's theorem tells us that *any* homogeneous equation involving only the values of G, m_1, m_2, r, and F can be put in the form $f(\pi_1, \pi_2) = 0$. For example, the law of gravitation is of this form, since it can be written as $\pi_1 \pi_2 - 1 = 0$. Note that we had to include G, even though it is a universal constant. *Everything* that can enter into the function must be included.

Two comments should be made about the mechanics of doing dimensional analysis. First, it is not always evident what should be included in the list of relevant physical variables and constants. The only sure guide is good intuition. Second, if you have had some linear algebra, you should recognize the procedure we went through to obtain π_1 and π_2: We found a basis for the two-dimensional subspace of R^5 that makes the exponents of M, L, and T in π equal to zero. Such a basis was given by $(a, b, c, d, e) = (1, 0, 2, -2, -1)$ and $(a, b, c, d, e) = (0, 1, -1, 0, 0)$. This procedure works in general: Find the exponents of M, L, and T in terms of the exponents of the exponents of the variables and constants appearing in π; then find a basis for the null space of these exponents. Each basis vector determines one of the dimensionless products π_i mentioned in the Buckingham pi theorem.

By formalizing the above idea we can obtain a proof of the Buckingham pi theorem. Here is a sketch for those who are familiar with linear algebra. Let $x_1, x_2, \ldots, x_k$ be the physical quantities we are studying. Define

$$f(x_1^{a_1} x_2^{a_2} \cdots x_k^{a_k}) = (a_1, a_2, \ldots, a_k).$$

This sets up a natural one-to-one correspondence between products of powers of x_i and the elements of R^k. We can replace each x_i by its dimensions and define another map d like f but this time into R^n. (Usually $n = 3$ for M, L, T.) Consider df^{-1}. It is a linear transformation from R^k to R^n. Let $b_1, \ldots, b_j$ be a basis for the null space and extend it to a basis $b_1, \ldots, b_k$ for R^k. Define $\pi_i = f^{-1}(b_i)$. We can express x_i as products of powers of the π_i, since the b_i form a basis for R^k. Hence any physical law expressed in terms of x_i can be expressed in terms of the π_i. For every $m > j$ there is a change in the units of measurement which changes π_m but leaves the other π_i unchanged. Since the laws of physics are assumed to be independent of the units of measurement, the law we are considering must be independent of π_m. Thus it depends only on $\pi_1, \ldots, \pi_j$, and these are all dimensionless since $b_1, \ldots, b_j$ lie in the null space of df^{-1}.

The Period of a Perfect Pendulum

Legend has it that Galileo's interest in motion began when he observed a hanging lamp in the Pisa cathedral swinging back and forth. This is an example of a pendulum. How fast does a pendulum swing? How does the period of the swing vary with the length? The weight? The angle of swing?

We consider a pendulum in which all the mass is concentrated at a distance l from a perfect pivot and there are no frictional forces. From observation or theory it can be determined that the motion of a frictionless pendulum is periodic with some period t. Since we want to derive a formula for t, it is our (only) endogenous variable.

What quantities should enter into such a formula? In other words, what are our exogenous variables? The length l of the pendulum, the mass m of the pendulum, the acceleration g due to gravity, and the maximum angle θ the pendulum makes with the vertical appear to form a complete list. (Since gravity is involved, you may wonder why G and the radius of the earth are not on the list. The only effect of gravity is to provide a force equal to mg acting on the pendulum. Both m and g are on our list.)

We now show that all dimensionless products can be formed from

$$\pi_1 = \frac{gt^2}{l}, \qquad \pi_2 = \theta.$$

The procedure is the same as the one we just used for the law of gravitation. We know the dimensions of l and t. The acceleration g has dimension LT^{-2}. Since an angle is measured by the ratio of arc length to radius, the angle θ is dimensionless. Thus the product $\pi = m^a g^b t^c l^d \theta^e$ has dimension $M^a L^{b+d} T^{c-2b}$. This vanishes if and only if $a = 0$, $b + d = 0$, and $c - 2b = 0$. It follows that we can choose b and e arbitrarily and that $a = 0$, $c = 2b$, and $d = -b$. We obtain π_1 from $(b, e) = (1, 0)$ and π_2 from $(b, e) = (0, 1)$. Note that m *does not appear*, because no other quantity has M in its dimension.

Since the period of a pendulum is a physical law, Buckingham's theorem applies. Solving $f(\pi_1, \pi_2) = 0$ for π_1 gives $\pi_1 = h(\pi_2)$ for some function h. Therefore

(15) $$\text{Period} = t = k(\theta) \sqrt{\frac{l}{g}},$$

where $k^2 = h$. The exact form of the function $k(\theta)$ must be determined by other means. It turns out to be an *elliptic integral* and is very nearly equal to 2π when θ is small.

Scale Models of Structures

Suppose you are an engineer and wish to study how a structure you've designed will hold up. Since theoretical analysis of a complicated structure is likely to be impossible, it is convenient to study a scale model. How should you design the model and how should the observations you make on it be translated into predictions about the real structure? We answer these questions here.

Unless they are greatly deformed, most structures can be reasonably approximated by assuming that they are built of materials that are *elastic* and *isotropic*. (These are technical terms.) The important physical consequence of this assumption is that, except for specifying shapes and forces, we need only two parameters to determine the changes in shape (called *deformations*). One is *Poisson's ratio* σ which is dimensionless. (It is the ratio of the percentage changes in the dimensions of a bar perpendicular and parallel to a compressive force.) The other is *Young's modulus E.* (It is the ratio of the compressive force per unit area to the percentage change in the parallel dimension.) The dimension of Young's modulus is $ML^{-1}T^{-2}$. The important thing is not how σ and E are defined, but rather the fact that *as far as deformations are concerned they are the only relevant inherent properties of the material(s).*

What are the relevant variables and physical constants? The endogenous variables are the deformations δ of the structure. Our structure has some characteristic length l by which we can relate all lengths of the scale model to those of the real structure. The specific gravity (weight per unit volume) γ may also be important. Weight per unit volume is density times acceleration due to gravity and so has dimension $ML^{-2}T^{-2}$. E and σ have already been mentioned. Finally there are the forces F which are loading the structure at various points.

Our list of relevant quantities is σ, γ, E, F, l, and δ. Actually, all of these except l should be subscripted to indicate that there may be several different materials and a variety of forces. All dimensionless products can be formed from the products

$$(16) \qquad\qquad \sigma_i, \qquad \frac{E_i}{E_1}, \qquad \frac{l\gamma_i}{E_1}, \qquad \frac{F_i}{(l^2 E_1)},$$

and δ_k/l. Each δ_k/l is determined by $\sigma_i, \gamma_i, E_i, F_i$, and l. It follows from Buckingham's theorem that δ_k/l is a function of the various products in (16). Therefore

> If the quantities in (16) are the same for the scale model and the real structure, all deformations will be scaled according to the scaling of l.

We must therefore keep σ_i the same for the materials in the scale model and the real structure. The easiest way to do this is to use the same materials in both cases. Then all the E_i and γ_i will be the same for the real structure and the scale model. From the third relation in (16) it follows that the two values of l must be the same, so the model is the same as the real structure.

How can we get around this? It is the *density* of a material that is constant; *the specific gravity γ varies with the gravitational field.* If we could adjust γ by changing the gravitational field, this would adjust l. How can we change the gravitational field? Since acceleration due to gravity is like any other acceleration, we can effectively increase "gravity" by using a centrifuge. This technique is actually used. Suppose the ratio of the scale model l to the real l is $1:r$. By the third expression in (16), the centrifuge must produce an acceleration r times that due to gravity. By the last expression, the model forces should be r^{-2} times the real forces. As a check of what we have been doing, let's look at a force F due to weight. It equals mg and so is proportional to γl^3. This changes by a factor of $rr^{-3} = r^{-2}$ as desired.

There is another approach to building scale models. The specific gravity of the materials is important only because it determines forces on the structure due to gravity. If we expand the list F_i to include these forces as well, we can neglect the specific gravities, hence the third type of ratio in (16). The fourth ratio tells us that forces must be proportional to l^2; however, if the gravitational field is unchanged (no centrifuge), the gravitational forces will vary as l^3. (Why?) To compensate for this the scale model of the structure can be loaded at various points with weights equal to the difference between these two quantities. This may make it necessary to measure forces at many points, but it eliminates the centrifuge. Without dimensional analysis these ideas for building scale models would have been hard to discover.

PROBLEMS

1. This problem relates to the pendulum model. We want to include frictional effects.

   (a) Suppose that the frictional force is due primarily to air and is proportional to v^2 with a constant of proportionality κ. The value of κ depends on the shape of the pendulum. Let τ be the time required for the pendulum to reach half its initial amplitude θ. Argue physically that v is determined by m, l, g, κ, θ, and the elapsed time. Show that

   $$\tau = \sqrt{\frac{l}{g}}\, f\!\left(\theta, \frac{\kappa l}{m}\right).$$

   (b) Deduce a similar result if the frictional force is proportional to v.

   (c) Using the results of (a) and (b), describe an experiment for deciding which (if either) of the assumptions about the dependence of the frictional force on v is correct. *Hint*: Consider a pendulum with a hollow weight which can be filled.

2. Why do stringed musical instruments have strings of different lengths and thicknesses? The fundamental frequencies of vibrations of strings of similar material depend primarily on length l, mass per unit length μ, and tension (force) F on the string.

(a) Derive the formula for the fundamental frequency ω for similar materials:

$$\omega \propto \frac{\sqrt{F/\mu}}{l}.$$

(b) In terms of the above result, explain the structure of a nylon string, six-string guitar. There are several structural constraints imposed on the instrument. Design and playing considerations dictate that the strings must be of the same length and cannot have either too large or too small a diameter, and impose upper and lower limits on the tensions in the strings. The frequency of the low string is only one-fourth the frequency of the high string. There is of course no need to explain these facts. The following properties of the guitar should be explained. When playing the guitar, different notes are obtained by using the fingers to shorten the length of various strings. Tuning is accomplished by adjusting the tension on the strings. The strings vary in thickness and in the material of which they are made. Roughly speaking there are three thicknesses ($T_1 < T_2 < T_3$) and two materials nylon (N) and steel-wrapped nylon (S). The strings, from highest frequency to lowest frequency, are NT_1, NT_2, NT_3, ST_1, ST_2, and ST_3.

(c) If you are familiar with the structure of another stringed instrument, interpret it as much as possible using the ideas in (a) and (b).

(d) One of my students (R. T. Oberndorf) collected data to check the formula in (a). He used guitar strings with an arrangement for changing the tension and the length. He found that for a given string ωl was constant to within the accuracy of his measurements when F was held fixed. The same was true for $\omega/\sqrt{F}$ with l held fixed. However, when he used various strings but fixed l and F, he found that $\omega\sqrt{\mu}$ was not constant. The largest deviations occurred with the thinnest strings and the highest tension, ω being higher than predicted. Suggest some possible explanations.

(e) Let's take the material of the string into account. We assume that the material is *elastic* and *isotropic*. Thus we need only consider Poisson's ratio σ and Young's modulus E. See the scale models of structures model in this section for a brief discussion of σ and E. Show that

$$\omega = K\left(\frac{El^2}{F}, \sigma\right)\frac{\sqrt{F/\mu}}{l}.$$

Use Oberndorf's result to show that K depends only on σ for guitar strings.

3. How long should you roast a turkey? Cookbooks usually give directions in the form: "Set the oven to T_0 degrees and allow n minutes per pound for cooking." For turkey, which can range in weight from about 7 pounds to about 30 pounds, a range of roasting times may be given. In this case, one cookbook recommends cooking for 15 to 25 minutes per pound, the longer time to be used for smaller birds. We study this in a problem adapted from S. J. Kline (1965).

(a) A piece of meat is cooked when its minimum internal temperature reaches a certain value dependent on the type of meat and the desired doneness. Let the cooking time t be the endogenous variable. Present an argument to show that the exogenous variables are the difference in temperature ΔT_m between the raw meat and the oven, the difference in temperature ΔT_c between the cooked meat and the oven, some characteristic dimension l of the meat, and some measure κ of the ability of the meat to conduct heat.

(b) The usual measure of ability to conduct heat is thermal conductivity which is the amount of energy crossing a unit cross-sectional area per second divided by the temperature gradient perpendicular to the area. Hence κ is measured in

$$\frac{\text{energy}/(\text{area} \times \text{time})}{\text{degrees}/\text{length}}.$$

The dimension of energy is ML^2T^{-2}. Temperature is measured in energy per unit volume. Determine the dependence of cooking time on the weight for similar pieces of meat for which ΔT_m and ΔT_c are the same.

(c) Discuss the accuracy of the cookbook rule. Comment on the rule for turkeys.

4. Waves seem to roll in at a beach in a regular fashion, but their speed seems to vary from place to place and, perhaps, from time to time. Why? Does something similar happen out at sea as well? We discuss wave motion in a perfect fluid; that is, a fluid with no viscosity or compression. Let the endogenous variable be the velocity v of a wave.

(a) Argue that the exogenous variables are acceleration g due to gravity, the density ρ of the liquid, the length λ of the wave, the height h of the wave, and the depth d of the liquid.

(b) When the height of a wave is small compared to its length, it is known that we can approximate the equations of motion by equations that do not contain h. Conclude that we can ignore h in this case.

(c) Show that

$$v = \sqrt{\lambda g} f\left(\frac{d}{\lambda}\right).$$

(d) Show that v is nearly proportional to $\sqrt{\lambda g}$ when d is large compared with λ. Thus wave speed at sea varies with the wavelength.

(e) Suppose we want to build a scale model to study the effect waves on the open ocean have on boats. How should everything be scaled? *Hint*: If all linear dimensions are scaled by a factor of r, what happens to the time it takes a wave to travel the length of the boat. You may also wish to refer back to the model dealing with scale models of structures.

(f) When d is small compared with λ, the bottom interferes with the wave, so that λ is practically irrelevant. Show that v is nearly proportional to $\sqrt{dg}$ in this case. The British government has used this result to obtain depth surveys in certain remote coastal areas. Two pictures were taken of the same region at slightly different times so that wave speed could be measured (R. Carson, 1961, p. 109).

CHAPTER 3

GRAPHICAL METHODS

3.1. USING GRAPHS IN MODELING

Graphs can be very useful in modeling if you are aware of their uses and limitations. Since many people expect either too much or too little from them, we discuss their uses and limitations before going into specific models.

People can take in an entire picture rather quickly and then deduce consequences by using their geometric intuition. It follows that graphs should be useful in conveying information. Those wonderful analog computers people carry in their skulls can rapidly locate certain patterns in visually presented data. One of the easiest to spot is a straight line. For this reason a variety of forms of graph paper (rectangular, polar, log–log, normal probability, etc.) are marketed so that plotted data will appear linear if the anticipated relationship exists.

Graphs are most useful in conveying qualitative relationships or approximate data which involve only a few variables. A graphical approach to a problem is most likely to be useful when not much information is available or when it is given in a rather imprecise form. Analytical methods are usually more appropriate when more precise information is available. In complex simulation models, graphs are frequently used to illustrate the qualitative behavior of several time varying endogenous variables simultaneously. This helps one obtain a qualitative feel for the behavior of a complicated simulation model.

So far we have talked about graphs primarily as a way of presenting data. Now let's consider some major roles graphs play in model formulation.

Since our imagination is limited to three dimensions, graphical representations of the interrelations of more than three variables are not directly useful. However, it is often possible to graph a function with most variables held fixed and then determine how the graph will change when one of the fixed variables is changed. This is the heart of the geometric approach to *comparative statics* which is discussed in Section 3.2. The differential calculus approach parallels the geometric arguments and provides a firm foundation for making statements when any number of variables is involved. The basic problem of comparative statics can be stated as follows: How does the equilibrium point of a system move when certain exogenous variables are changed? For example, how will the output of a firm be affected by a higher tax rate?

Graphical methods are also useful in studying *stability questions*. The analytical treatment of local and global stability theory is not easy. Therefore it is desirable to use graphical methods whenever possible to suggest and perhaps prove results. Section 3.3 touches on this approach. For a treatment of the problems of stability theory from an analytical viewpoint see Chapter 9.

As a glance at the figures in this chapter shows, the *intersections* of curves are of major importance in comparative statics. This is because they determine the equilibrium points. A subtler observation is that *slopes* of curves play a central role in stability questions. The slope of a curve is a rate, and rates play a crucial role in stability theory.

Finally, graphical arguments are useful in *optimization* problems—especially if the model is not quantitative. Since this straddles Chapters 3 and 4, I've decided to put it in Section 4.2.

3.2. COMPARATIVE STATICS

The Nuclear Missile Arms Race

The United States and the U.S.S.R. both feel that they require a certain minimum number of intercontinental ballistic missiles (ICBMs) to avoid "nuclear blackmail." The idea is to ensure that enough missiles will survive a sneak attack so that "unacceptable damage" can be inflicted on the attacker. Given this philosophy, it is claimed by some and denied by others that the introduction of antiballistic missiles (ABMs) and/or multiple warheads on each missile (MIRVs) will cause both nations to increase their stock of missiles. Is this true? What about making missiles less vulnerable to attack by hardening silos or building missile firing submarines? The wrong answers to these questions could have drastic consequences. Who is right?

Obviously we cannot hope to settle the debate. However, a simple graphical model can shed some light on the problems involved and hopefully help lead to more intelligent debate. The following discussion is adapted from T. L. Saaty (1968, pp. 22–25).

We deal with two countries which we call country 1 and country 2. Let x and y be the number of missiles possessed by countries 1 and 2, respectively. We treat x and y as real numbers. Of course they are actually integers; but since they are large, the relative errors introduced by treating them as real numbers will be small; for example, the percentage difference between 500 and 500.5 is quite small. For the time being we assume that all missiles are the same and are equally protected. From the above discussion it follows that there exist continuous, increasing functions f and g such that country 1 feels safe if and only if $x > f(y)$, and country 2 feels safe if and only if $y > g(x)$. These functions are plotted in Figure 1. The shaded region is the area in which armaments are stable, since both countries feel they have sufficient weapons to prevent a sneak attack. We consider questions such as: Does such a region actually exist? What effect do such things as ABMs, MIRVs, and so on, have on the point $A = (x_m, y_m)$?

First we show that the solid curves in Figure 1 are qualitatively correct. Let's look at things from the point of view of country 1. A certain number of missiles x_0 is needed to inflict what is considered unacceptable damage on country 2. When country 2 has no missiles, country 1 requires x_0.

We show that for any $r > 0$ the curve $x = f(y)$ crosses the line $y = rx$. It suffices to show that there is a function $x(r)$ such that, whenever $x \geq x(r)$

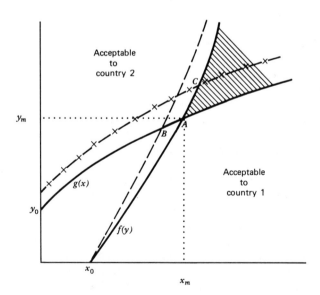

Figure 1 Country 1 introduces ABMs. A = initial status (shaded area stable); B = country 1 protects its missiles; C = country 1 protects its cities. Axes show number of missiles.

and $y = rx$, country 1 believes that it has enough missiles so that the number surviving a sneak attack by country 2 will be able to inflict unacceptable damage on country 2. In other words, country 1 wants to be practically certain of at least x_0 of its missiles surviving a sneak attack by country 2. Suppose that $y = rx$. To destroy the most missiles, country 2 should aim about r missiles at each of country 1's missiles. Since a warhead may fail to reach and destroy its target, there is some probability, $p(r) > 0$, that a given missile belonging to country i will survive a sneak attack. Thus country 1 can expect $xp(r)$ missiles to survive. For large enough $x = x(r)$, this will exceed x_0 by an amount large enough to allow for uncertainties. This completes the proof that the curves intersect. Thus the curve $x = f(y)$ starts at $(x_0, 0)$ and curves upward with a slope increasing to ∞. By a symmetry argument, $y = g(x)$ has the form shown, with a slope decreasing to 0. Two such curves meet at exactly one point which we call (x_m, y_m), the minimum stable values for x and y.

This analysis applies to all the situations discussed below, so there is always a unique minimum stable point. We want to know how its position compares with (x_m, y_m).

Suppose the missiles of country 1 are made less vulnerable to sneak attack by the use of hardened silos, ABM protection, or some other means. This increases $p(r)$, the probability that any given missile belonging to country 1 will survive a sneak attack. Hence the curve $f(y)$ moves to the left with the point x_0 fixed. The shape of the curve is altered somewhat in the process. The new curve is shown dashed in Figure 1. We can see that *both countries require fewer missiles for stability*.

Suppose that country 1 protects its cities by some device such as ABMs. Country 2 now requires more than y_0 missiles to inflict unacceptable destruction on country 1. Thus the curve $g(x)$ moves upward as shown by the $\times - \times - \times$ curve in Figure 1. *Both countries require more missiles for stability*.

What happens if multiple warheads are installed? This situation is more complicated than the previous two. Suppose country 1 replaces the single warheads on each of its missiles with N warheads. It will then require that fewer of its missiles survive a sneak attack. (The number required is about x_0/N.) Thus $x = f(y)$ moves to the left as in Figure 2. Country 2 will be faced with N times as many warheads in a sneak attack, so from its point of view the scale of the x axis has changed by about a factor of N, as shown in Figure 2. It appears that country 2 will require more missiles, and country 1 will require fewer; however, this depends on the detailed shape of the curves. Therefore probabilistic models should be used instead of, or in conjunction with, graphical ones. This would require us to make more precise assumptions regarding the capabilities of the missiles, so we do not go into it here.

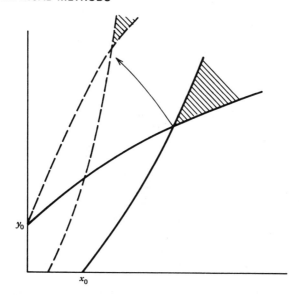

Figure 2   Country 1 introduces MIRVs. Axes show number of missiles.

It seems unreasonable to assume that country 2 will not also develop and deploy multiple warheads if country 1 does. Therefore we should analyze the situation in which both countries deploy multiple warheads. There are two conflicting effects:

1.   Since the axes measure *missiles*, the points $[f(0), 0]$ and $[0, g(0)]$ will move toward the origin, tending to decrease (x_m, y_m).
2. $f(y)$ becomes more horizontal and $g(x)$ becomes more vertical, tending to increase (x_m, y_m).

We cannot decide without further information which effect will dominate. T. L. Saaty (1968, p. 24) presents an analytical model which leads to the conclusion that both countries will require many more missiles.

In the above discussion, we assumed that all missiles were the same. This is unrealistic. If we drop this assumption, each country will change its strategy by aiming different numbers of missiles at the various enemy missiles. Of these, some targeting makes the expected surviving firepower a minimum. This targeting gives the curves for Figure 1, and the analysis proceeds as before.

You may be interested in the article by K. Tsipis (1975a) which contains a discussion of the technology behind ultraaccurate MIRVs.

Biogeography: Diversity of Species on Islands

The diversity of species varies considerably from place to place, even when the habitats appear to be the same. Conservationists have argued that the size of a region is important for diversity, and so they often favor a few large wilderness areas rather than many tiny ones. The subject is far from understood. We study one corner of it briefly.

The world is broken into patches of differing habitats. Often a habitat a species finds acceptable is surrounded by a large expanse of unacceptable territory. Examples are alpine meadows, farm woodlots, lakes, game preserves, and islands. The following discussion is confined to islands; however, most of the ideas and results apply to other types of isolated habitats. The material is adapted from R. H. MacArthur and E. O. Wilson (1967, Ch. 3) which treats the subject in much greater depth.

Studies have indicated that the size of an island is an important factor in determining the number of species the island is likely to contain. Also, islands closer to the mainland tend to contain a greater variety of species than more isolated islands. It seems reasonable that the effects of migration of species and extinction of species (on islands) can account for this. We develop this idea and briefly consider some of its consequences.

A species can become established on an island only by migrating to it and prospering there. An organism migrates by flying, being carried, drifting on currents, and so on. Since a population on an island is relatively small, it can die out because of random variations in the environment. As a result we expect the list of species present on an island to change much faster than the list of species present on the mainland.

This is somewhat vague. Does a flock of migrating birds that stops on the island for a day or a season become established and then die out? Even if a species "intends" to stay on the island, we are still faced with the problem of what we mean by "become established"—if the island is too small to support a large population, the species will always be on the verge of extinction. When is a species established in this case? Since we are dealing with a fairly crude model, we can afford to ignore these problems. A more refined model would have to come to grips with them.

If we *completely* understood all the aspects of the situation (e.g., the biological, geographical, and meteorological), we could determine the probability of a particular species composition being present on the island at a given future time. These would be tremendously complex calculations involving vast quantities of data, and this approach would be hopeless.

Let's combine practically all the endogenous variables into one measure: the total number of species present on the island. It seems reasonable to suppose that this should vary around some average number of species in a steady state situation. We discuss this average. For a discussion of transient behavior see R. H. MacArthur and E. O. Wilson (1967, Ch. 3) or E. O. Wilson and W. H. Bossert (1971, Ch. 4).

When the number of species present on the island is *in equilibrium, migration and extinction cancel out numerically*; that is, the rate of migration of new species to the island equals the rate of extinction of species already on the island. These rates depend in a complicated way on the species present, the season, and many other factors. If we regard a year as a short period of time, seasonality will present no problem. In this sort of crude averaging over many species, which species are actually present probably doesn't matter much. Therefore it makes sense to talk about rates in a crude way independent of which species are actually present on the island.

In Figure 3 are plotted the number of species N on the island versus the migration and extinction rates. The two smaller graphs illustrate the effect of distance from the mainland and the effect of island size. We discuss the reasons for the shapes and positions of the curves.

Let's consider the extinction rate curves. When more species are present on the island, the chances that at least one species will become extinct in a given time are greater. Hence the extinction rate curves have a positive slope. Since extinction rates depend only on the island and the species present, the extinction curve is not affected by the distance from the mainland. However, we can expect that a species is more likely to die out on a small island because the lack of space keeps the population lower. Thus the extinction rate curves shift upward as the islands become smaller.

Why do the migration rate curves have a negative slope? The migration rate relates to species *not present on the island*. The greater the diversity on the island, the smaller the pool of potential migrating species on the mainland. Hence the chances of migration decrease as the number of species on the island increases. Migration rates depend on the distance of the island from the mainland and on the size of the island. The rates decrease with distance, because any given organism is less likely to reach the island. The rates increase with island size because (1) an organism has a larger land area as a target and (2) an organism is more likely to be able to establish itself on a larger island.

It follows from Figure 3 that the number of species present increases with island size and decreases with distance from the mainland. This is not so surprising, since we practically put these results in as initial assumptions.

We can say something about species turnover by looking at the graphs a bit more. Note that the equilibrium extinction rate (which equals the

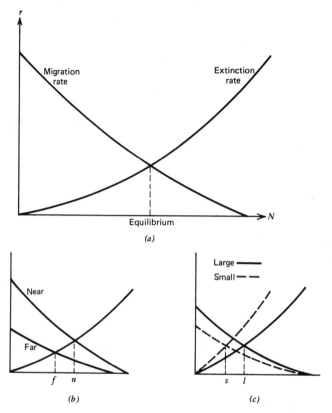

Figure 3 Migration and extinction curves for islands. (*a*) Typical curves. (*b*) Effect of distance. (*c*) Effect of size.

equilibrium migration rate) is greater for near islands than for far islands. Hence the species composition for two islands of equal size should change more rapidly on the island closer to shore. If the effect of island size on migration rate is not too great, we can similarly conclude that the species composition changes faster on small islands than on large islands. Since small islands have fewer species at equilibrium than large islands, this effect should be quite noticeable.

There is some data supporting the conclusions that species turnover is relatively and absolutely more rapid on smaller islands. R. H. MacArthur and E. O. Wilson (1967, pp. 52–54) discuss the results of two botanical surveys of some small islands off the Florida Keys. The first survey (1904) was conducted quickly and so may be incomplete. Since the 1916 survey was quite complete, the species present in 1904 and absent in 1916 give some measure of the turnover rate. Unfortunately the data involve only six islands, two of which are very small.

R. H. MacArthur and E. O. Wilson (1967, pp. 55-60) also report on some results of R. Patrick. She suspended glass slides in a spring in Pennsylvania and counted the number of diatoms of various species that were present. The glass slides can be thought of as islands. Four experiments were done two times each: A glass slide with an area of either 12 or 25 square millimeters was placed in the water for either 1 or 2 weeks. The slides submerged for 1 week had more species present than those submerged for 2 weeks. We can explain this apparent contradiction by observing that as a barren area becomes more populated the interaction between species may cause extinction. This was not allowed for in our model. Clearly care must be taken in modeling islands that are far from equilibrium. Because of this, we do not consider the 1 week data further. We can check out two predictions using the 2 week data:

1. Larger area implies more species: The smaller slides had 24 and 21 species, and the larger had 29 and 28.
2. Smaller area implies a higher migration rate: The migration rate may be reflected somewhat in the differences in the species composition of the slides. (Why?) Seven species appeared on one but not both of the smaller slides. For the larger slides the number was one.

Theory of the Firm

You are the manager of a firm which produces, among many other items, "zowies." How can you decide on a level of production? The price of the main raw material for your zowies is going to increase. Perhaps you can pass some of the cost on to your customers. How much? Can you pass on enough to make it worthwhile to continue manufacturing zowies? Quantitative results are hard to obtain because data collection is extremely difficult; however, we can obtain a qualitative picture of the situation fairly easily.

In the usual theory of the firm it is assumed that the manager of the firm has complete information, that his decisions are carried out, and that he acts so as to maximize the profits of the firm. There is an ongoing debate about the usefulness of these assumptions, but we don't want to get into that here. If you are interested in the subject see R. M. Cyert and J. G. March (1963, pp. 5-16) for a discussion of both sides of the question. In addition to the above assumptions, we generally assume (as is often done in economic theory) that the functions with which we are dealing are well behaved; that is, they are continuous and usually differentiable.

The theory of the firm is discussed in most textbooks on mathematical economics. There also exist books devoted exclusively to the topic, such as K. J. Cohen and R. M. Cyert (1965). Consult such sources if you wish to see the ideas in this example developed further.

For simplicity we assume that the firm produces only one product, so that we can speak unambiguously of the level of production. It is measured in units per time period, where a time period can be a day, a month, or any other convenient interval. We want to find a way to determine the level of production, so that we can discuss the influence of changing costs and prices on the production level.

Suppose the production of the firm is at some equilibrium level. Since profits are being maximized, the additional cost that would be incurred in raising production slightly is equal to the additional gross income that would be obtained by marketing these additional units of the product. You should convince yourself that this is simply a restatement of the calculus theorem that the function

$$\text{Total gross income} - \text{Total cost}$$

has maxima and minima where its derivative vanishes.

The additional cost required to produce one additional unit is called the *marginal cost*, and the additional income is called the *marginal income*. In general, both marginal cost and marginal income are functions of the level of production. We have shown that marginal cost equals marginal income at equilibrium. This equality could imply that the profit is a minimum instead of a maximum. How can we distinguish one from the other? If we move away from the equilibrium, profits must decrease. Thus the marginal cost curve must lie above the marginal income curve for higher production levels and below it for lower production levels. This is shown in Figure 4 where the horizontal axis is the quantity produced per unit time. This is the basic result with which we work.

Although marginal cost and marginal income may seem to be straightforward concepts, they can be a bit fuzzy. During a short period of time (the short term), wages and the cost of raw materials are *fixed costs*, because they have been contracted for; consequently, they do not enter into marginal calculations. From a slightly longer point of view, they are both variable costs and so enter into the marginal costs. Equipment depreciation is a fixed cost; but maintenance, fuel, and replacement costs enter into the marginal calculations. Since our marginal curves vary with how long a view we take, the optimum level of output may depend on the length of time we want to

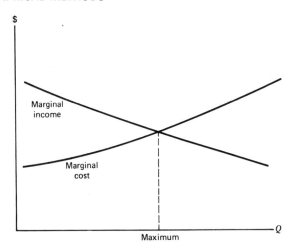

Figure 4 Marginal cost and income curves. Axes show quantity produced per unit time and dollars per unit time.

consider. In the following discussion we make the vague assumption that the manager is concerned with the firm's profits over a reasonably long time interval. As long as we don't try to make any detailed applications, we can afford to be vague.

What effect will taxation have on production? If the firm is required to pay a lump sum tax independent of production (e.g., a property tax), the marginal curves will not be affected. Hence *the production level will be unchanged*. If the firm is required to pay a tax that depends on the level of production (e.g., an income tax or a value-added tax), the result will depend on whether or not the tax is passed on to the consumer. If it is not passed on, the marginal cost curve will rise. We have shown that the marginal cost curve intersects the marginal income curve from below at a maximum. It follows that the new intersection will be to the left of the old one. Therefore *the production level will decrease*. What will happen if the tax is passed on to the consumer? In this case both marginal curves will move upward by an amount equal to the tax per unit of production, and the production level will be unchanged.

The above result on taxation can be generalized considerably, and we can state another result on the income side:

The optimum level of production moves in the opposite direction from the marginal cost and moves in the same direction as the marginal income.

Convince yourself that this is true by giving a graphical argument. Suppose the price of raw materials increases. This raises the marginal cost, so the production level tends to decrease. Decreased production may cause consumers to drive up the cost (per unit) of the product, thereby increasing the producer's marginal income. Consequently the level of production will rise. Since the product now costs more, the amount purchased by consumers will probably be less. Thus the increase in cost will not be quite enough to push production back to its original level. We discuss this in terms of supply and demand curves.

In industries where the number of firms is large, it is reasonable to suppose that the price per unit of product does not depend on the amount any single firm produces. In this case the marginal income curve is horizontal. The marginal cost curve is then the *supply curve* for the firm's product, since at a price p the firm produces the quantity Q at which the marginal cost equals p. Since the marginal income curve is horizontal, our earlier discussion shows that the supply curve must have a positive slope to ensure stability. This agrees with the intuitive notion that higher selling prices lead to greater production.

The *demand curve* is the amount of the product that will be purchased at a given price. Usually demand falls as price increases. Figure 5 shows typical supply and demand curves. At equilibrium, the quantity purchased must equal the quantity sold. Hence the intersection of the supply and demand curves gives the equilibrium values of price and quantity.

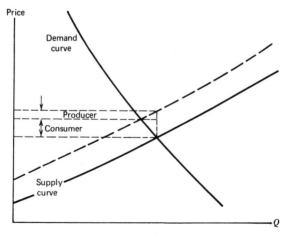

Figure 5 Supply and demand curves. Increased marginal costs shift supply curve upward to dashed position.

From Figure 5 we can see how much of the increased marginal costs will be passed on to the consumer. The dashed curve shows the supply curve (marginal cost curve) after the marginal costs have increased. The flatter the demand curve, the greater the fraction of the increase the producer must absorb. What does a flat demand curve mean? It indicates that consumer buying patterns are very sensitive to price. Thus, if consumer buying patterns are insensitive to price, you can pass most of your increased expenses on to the consumer.

What about the theory of a firm that produces several products? It is better to study such a situation using tools from calculus. However, our graphical analysis indicates the sort of results we can expect to find in this case.

PROBLEMS

Problems 1 to 5 deal with the arms race model.

1. Suppose that both countries install N warheads in each missile and that the new warheads are as effective as the old ones. Show that both countries will require more *warheads*.

2. Suppose a country is able to retarget missiles in flight so as to aim for missiles that previous warheads have failed to destroy. Discuss the effect.

3. Various criteria have been used to evaluate proposed changes in missile systems. Try to evaluate the changes discussed in the text and the problems on the basis of (a) economics (cost) and (b) amount of radio-activity released in the event of a war.

4. There are aspects of the armaments race that become important only when a country is not as heavily armed as the United States and the U.S.S.R. When a country is just developing a nuclear strike force, it may be able to inflict heavy damage with a first strike but may be in-capable of a retaliatory strike.

    (a) Develop a model and use it to explain "preventive war." Can you apply the model to the People's Republic of China?

    (b) Can you model the early years of the missile race?

    This is a rather unclear area, so class discussion may lead to a variety of ideas. You may wish to consult M. D. Intriligator (1973).

5. The United States and the U.S.S.R. signed an arms limitation agreement in May 1972. The number of offensive *missiles* allowed each country is limited, with a trade-off formula for land-based versus submarine-

based missiles. There is no limitation on the use of multiple warheads or on improving missile technology. Each country is limited to two ABM sites of 100 missiles each. One site is for protection of the capital city and the other for protection of an ICBM site.

(a) Discuss this agreement in light of the models presented here. Include any relevant later agreements in the discussion. Politics is more complicated than our simple model, so you will have to weigh various factors that might affect the model's validity.

(b) How can the model be improved to help in answering (a)?

6. Will a group of small islands have more or fewer species per island than an isolated small island? Assume that all the islands are about the same distance from the mainland and the same size.

7. Discuss what happens in the model dealing with the theory of the firm if the marginal cost curve does not intersect the marginal income curve.

8. In the short term, ordinary wages are a fixed cost and overtime wages are a marginal cost.

(a) Explain the previous statement.

(b) Show that the marginal cost curve has a discontinuity at the level of production corresponding to full usage of labor without overtime.

(c) What effect will this have on the results developed in the model of production by a firm?

3.3. STABILITY QUESTIONS

Cobweb Models in Economics

We consider the dynamics of supply and demand when there is a fairly constant time lag in production as, for example, in agriculture. It has been observed that there are fairly regular price fluctuations in such situations. This situation was studied by economists in the 1920s and 1930s. The problem contrasts sharply with the theory of the firm in Section 3.2, where we ignored time. The following discussion is adapted from M. Ezekiel (1937/8).

When a commodity is marketed, the selling price is determined by the *demand curve*. This price is one of the factors producers use in determining how to alter production. In a "pure" situation, they produce the amount on the *supply* curve that corresponds to the present price. (Supply and demand curves are discussed more fully in the theory of the firm model in Section 3.2. There we were interested in the intersection point of the curves.) Thus (see

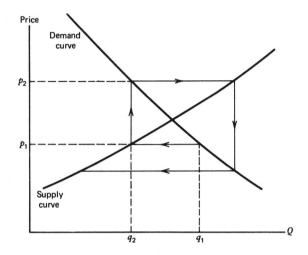

Figure 6 The cobweb model.

Figure 6), if the amount of potatoes produced in year 1 is q_1, the price per bushel will be p_1. As a result, farmers will decide to produce the amount q_2 in year 2, the market will set a price p_2 per bushel for this crop, and so on. Because of the picture, this idea is referred to as the *cobweb theorem*. In practice one does not know the supply and demand curves, but the above model predicts that the demand curve can be obtained by plotting (q_n, p_n) and the supply curve by plotting (q_n, p_{n-1}).

How realistic is this model? The existence of a supply curve assumes that producers can control output perfectly. This is not true in the agricultural sector where weather is very important, but it may be a reasonable approximation. If the supply and demand curves move erratically, the model will be upset. Changes in prices for other goods the supplier may produce, sudden changes in demand (e.g., the sale of wheat by the United States to the U.S.S.R. in 1972), and sudden changes in supply (e.g., crop blights) may cause this to happen. If the suppliers have some understanding of price fluctuations, they will not raise production levels much in spite of higher prices. However, this does not wreck the model. In this case the supply curve will be nearly independent of price near the equilibrium price, but the model will still apply. It predicts small fluctuations in supply and a rapid approach to stability. Plot this.

Ezekiel presented the material on U.S. potato production contained in Table 1. He obtained it from the Bureau of Agricultural Economics.

Table 1   Potato Production in the United States

Year	10^4 acres	Bushels/acre	10^6 bushels	Farm price	Deflated price
1921	360	90	325	114	121
1922	395	106	419	69	68
1923	338	108	366	92	93
1924	311	124	384	71	71
1925	281	106	296	166	162
1926	281	114	322	136	140
1927	318	116	370	108	113
1928	350	122	427	57	59
1929	302	110	332	132	142
1930	310	110	341	92	116
1931	347	111	384	46	68
1932	355	106	376	39	62
1933	341	100	342	82	114
1934	360	113	406	45	57
1935	355	109	386	60	74
1936	306	108	330	111	132

Discuss what should be used as "quantity" and what should be used as "price" in a cobweb plot and construct the plot. Should the model be modified because the yield per acre is not constant? What about the effect of population growth during the 15 year period? What about the effect of the Depression? Clearly there is a lot of noise (i.e., disturbances we can't hope to take into account in a simple model) in the data. Thus we should see if the data fit the model better than a random set of data would. Can you propose a method for doing this?

From the supply and demand curves near equilibrium it is easy to make a prediction concerning stability. If the negative of the demand curve's slope exceeds the slope of the supply curve, there will be instability; if it is less, stability. Convince yourself of this. Demand for some agricultural products is rather inflexible. When production is sensitive to price, the model predicts instability. The government can attempt to eliminate this by controlling production or prices. The former causes the supply curve to become vertical (or nearly so) above (and/or below) certain ranges of quantity. This keeps the instability from growing further. (Draw a graph to convince yourself.) What is the effect of price control?

For a further discussion of cobweb models see N. S. Buchanan (1939) and, for a recent generalization, M. S. Mudahar and R. H. Day (1974).

Phase Planes

The previous model dealt with the stability of a difference equation. A similar procedure is used for differential equations. This requires the notion of a *phase plane*, which is also used in Chapter 9. Suppose we are dealing with the two equations

(17) $$x' = f(x, y), \qquad y' = g(x, y).$$

At each point (x, y) in the $x - y$ plane we can plot a vector proportional to (x', y'). This is called the *direction field* of (17). To graph a solution of (17) we then start at an initial point and follow a path parallel to the direction field. (Since the direction field varies from point to point, the path is usually curved.) The speed is determined by the magnitude of the vector tangent to the path at that point. If we start at a point with $f = g = 0$, we will not move from it. Such points are called *equilibrium points*.

Since we have only crude information about f and g, our phase plane diagrams cannot be this detailed. To answer stability questions it is often sufficient to plot the two curves $f = 0$ and $g = 0$ and indicate roughly the vectors (x', y') in the neighborhood of these curves. The intersections of the curves are the equilibrium points of (17). The curve $f = 0$ divides space into two regions such that $x' > 0$ in one and $x' < 0$ in the other. If you determine which region is which for $f = 0$, and likewise for $g = 0$, the rest will be easy. The vectors cross $f = 0$ vertically, and the direction will be upward if and only if $g > 0$. Similarly, they cross $g = 0$ horizontally, and the direction will be rightward if and only if $f > 0$. See Figure 7 on page 63 for an example. In plotting $f = 0$ and $g = 0$, it is helpful to determine the slopes of the curves. This can be done by implicit differentiation: For $f = 0$,

$$\frac{dy}{dx} = - \frac{\partial f/\partial x}{\partial f/\partial y},$$

and similarly for $g = 0$. It is important to remember that the partial derivatives for the slope of $f = 0$ are evaluated at values of x and y at which x is at equilibrium; that is, $x' = 0$. (This is important in determining the sign of $\partial f/\partial x$ in Problem 4a.) The partial derivatives also help decide which region corresponds to $f > 0$ and which to $f < 0$: $f > 0$ to the right of (or above) $f = 0$ if and only if $\partial f/\partial x > 0$ (or $\partial f/\partial y > 0$).

Small-Group Dynamics

You wish to set up a local committee to help elect a candidate to office. What keeps a group together and working? Does more work improve a task-oriented group or harm it? Very little mathematical modeling has been done in this area and, unfortunately, the following is rather crude and lacking in practical advice.

We want to study the stability and comparative statics of a group which has a required activity imposed from the outside (a task). The model is taken from H. Simon (1952), who based it on a nonmathematical model proposed by G. C. Homans (1950).

There are four basic functions of time:

$I(t)$, the intensity of *interaction* among the group members.

$F(t)$, the level of *friendliness* among the group members.

$A(t)$, the amount of *activity* within the group.

$E(t)$, the amount of activity imposed on the group by the *external environment*.

The variables can be treated as averages over all group members or as some overall measure for the entire group. We regard I, F, and A as endogenous variables and E as an exogenous variable which we generally treat as being constant.

To make the concepts more concrete, let's consider an example. The imposed activity E is the laying in of firewood. The group may be engaged in this for wages, or they may be friends preparing for winter. The various activities A include locating wood sources, sawing logs, stacking logs, and setting up a football pool. Note that some activities may not be directed toward the externally imposed task. G. C. Homans (1950, p. 101) says, "By our definition interaction takes place when the action of one man sets off the action of another." "Action" here refers to activity, so that activity is required for interaction, but not conversely—a person can work alone. The many situations in our example that involve interaction include discussing where to obtain wood, working opposite ends of a saw while cutting logs, passing wood from one person to another in stacking, and conversing idly. Some of the interaction is necessary, but a lot of it can be reduced considerably. The same is true of activity, as any efficiency expert knows; however, this may involve changes in habit patterns and so require more time.

There are three relations on which the model is based:

1. $I(t)$ depends on $A(t)$ and $F(t)$ in such a way that it increases if either A or F does. The adjustment is practically instantaneous.
2. $F(t)$ depends on $I(t)$. It tends to increase when it is too low for the present level of interaction and to decrease if there is not enough interaction to sustain its present level. This adjustment requires time, and the rate of adjustment is greater when the discrepancy between present and equilibrium levels is greater.

3. $A(t)$ depends on $F(t)$ and $E(t)$. It tends to increase when it is too low for the present level of F or E and to decrease when it is too high. This adjustment requires time, and the rate of adjustment is greater when the discrepancy between present and equilibrium levels is greater.

Criticize the assumptions.

These assumptions can be turned into equations:

(18a) $$I(t) = r(A, F), \qquad \frac{\partial r}{\partial A} > 0, \quad \frac{\partial r}{\partial F} > 0,$$

(18b) $$F'(t) = s(I, F) \qquad \frac{\partial s}{\partial I} > 0, \quad \frac{\partial s}{\partial F} < 0,$$

(18c) $$A'(t) = \psi(A, F : E) \qquad \frac{\partial \psi}{\partial A} < 0, \quad \frac{\partial \psi}{\partial F} > 0, \quad \frac{\partial \psi}{\partial E} > 0.$$

The reasoning behind $\partial s/\partial F < 0$ and $\partial \psi/\partial A < 0$ deserves an explanation. The same idea applies to both cases. Let's consider ψ. If A, F, and E are at some level, $\psi = A'$ will be determined. If we now increase A, we will either reduce the pressure for A to increase (if $\psi > 0$) or increase the pressure for A to decrease (if $\psi < 0$). In either case $\partial \psi/\partial A < 0$.

By substituting (18a) into (18b) we obtain

(19)
$$F'(t) = \varphi(A, F), \qquad \frac{\partial \varphi}{\partial A} = \frac{\partial s}{\partial I} \bigg/ \frac{\partial r}{\partial A} > 0$$

$$\frac{\partial \varphi}{\partial F} = \frac{\partial s}{\partial F} + \frac{\partial s}{\partial I} \bigg/ \frac{\partial r}{\partial F}.$$

This equation says that a high level of A tends to cause F to increase. The effect of a high level of F is ambiguous: It may tend to cause F to increase or decrease. The statement that $\partial \varphi/\partial F > 0$ can be interpreted as: "The greater the friendliness, the faster it tends to increase (or the slower it tends to decrease, if it is decreasing)." While this may be true at some points in the A-F plane, it is unlikely to be true when F is large because of limits on friendliness. *We assume that $\partial \varphi/\partial F < 0$ everywhere.* The curves $\psi = 0$ and $\varphi = 0$ are plotted in Figure 7. The slope of the curves is positive, since, for example, on the curve $\psi = 0$, $dF/dA = -(\partial \psi/\partial A)/(\partial \psi/\partial F) > 0$. The slope of the $\psi = 0$ curve is increasing, because we assume a saturation effect: When A and F are both large and $A' = 0$, a fairly large increase in F is required to balance a small increase in A. In other words, the group tends to resist increases in activity more when it is already quite active. Discuss the curve $\varphi = 0$. Verify the general shape of the direction field shown in

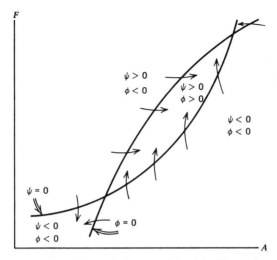

F

$\psi > 0$
$\phi < 0$

$\psi > 0$
$\phi > 0$

$\psi < 0$
$\phi < 0$

$\psi = 0$

$\psi < 0$
$\phi < 0$

$\phi = 0$

A

Figure 7 Dynamics in the activity–friendliness plane.

the figure. It can be seen that the upper equilibrium point is stable and that the lower one is unstable.

We now consider the effect of changing E. We have

$$\Delta\psi \approx \frac{\partial\psi}{\partial A}\,\Delta A + \frac{\partial\psi}{\partial F}\,\Delta F + \frac{\partial\psi}{\partial E}\,\Delta E.$$

Since $\Delta\psi = 0$ on the curve $\psi = 0$, it follows from (18c) that, when $\Delta A = 0$, ΔE and ΔF have opposite signs. Thus the $\psi = 0$ curve moves downward as E increases. Hence

The equilibrium levels of A and F are increasing functions of E.

If the $\psi = 0$ curve moves sufficiently far up, it will no longer intersect the $\varphi = 0$ curve, and so there will be no equilibrium point. In this case the group will not continue to exist. Consequently it is possible that a group will break up if externally imposed activity falls below a certain level.

PROBLEMS

1. Discuss modifications of the cobweb model when there is a time lag of more than 1 year in production, for example, raising hogs. The prices for hogs and corn (the principal feed for hogs) oscillate, and there is a fairly good correlation when they are offset a bit. Explain.

2. The demand for new graduates in various fields fluctuates. How should your department adapt its graduate program to help stabilize the situation? This problem is purposely very vague in hopes of generating a discussion based on reasonable models. Don't forget that feasibility is important. Engineering departments have gone through at least two cycles.

3. Discuss the group interaction model when $\partial \varphi / \partial F > 0$ for small F.

4. Suppose that two species are in competition. Let the number of members of the first species in the population be x and the number of the second be y. Assume that the environment if fairly constant.

    (a) Show that it is reasonable biologically to suppose that there exists a curve $y = r(x)$ of *negative* slope such that species 1 increases if and only if (x, y) lies below the curve.
    (b) State the corresponding assumption for species 2.
    (c) Show that the equilibrium points are the intersection points of the curves, the point $(0, 0)$, the point $(f(0), 0)$, and the corresponding point for species 2.
    (d) Determine the stability of the various possible equilibria.

5. You are called upon to advise an underdeveloped country on methods for increasing per-capita income. This problem briefly considers two difficulties you may encounter. It is an economics theory result that per-capita income is greater when accumulated capital per capita is greater. The idea is that, under suitable assumptions, since more capital is available it is used to help improve production. Do you think this applies to underdeveloped countries? What happens if capital is invested abroad or foreign capital is brought in? Let's assume that the economies theory result still applies. By definition, the capital accumulated in a year equals income (i.e., production) minus consumption.

    (a) Fractional rates of growth are defined in the same way as net growth rates in biology: $x'(t)/x(t)$. We denote the fractional rate of growth of x by x^*. Let K stand for total capital and P for total population. Show that per-capita income is increasing if and only if $K^* > P^*$.
    (b) One could suppose that P^* and K^* depend on per-capita income. Argue this point. Supposing it to be true, plot P^* and K^* as functions of per-capita income and show that intersections of the curves correspond to equilibria. How can you determine stability?
    (c) In each of the following cases, discuss the shape of the K^* and P^* curves near the given income level and use (b) to explain why these effects can keep per-capita income from increasing.

(*i*)   Rising expectations: At a certain income level, savings decrease because people try to mimic more affluent societies.

(*ii*)   Population explosion: At a certain income level, improved sanitation and diet reduce the death rate, but the birth rate takes much longer to fall because it is the result of custom.

(*d*)   That's the background for showing the ministers of the country some of the problems they face and what is going on. Now, advise them.

See P. A. Neher (1971, Ch. 8) and J. C. G. Boot (1967, Ch. 11) for further discussion.

CHAPTER 4

BASIC OPTIMIZATION

Determining what must be maximized (or minimized) is usually a major problem in formulating an optimization model. For example, the theory of the firm assumes that managers behave so as to maximize profit; but it has been suggested in recent years that they maximize a *utility function*, which includes size of staff and other items in addition to profit. Another example is provided by *time sharing algorithms* for computers. (A time sharing algorithm is an algorithm used by a computer to decide which of many waiting jobs to run and how long to let it run before interrupting it temporarily to run other jobs.) What should be minimized? Among the myriad of possible functions are

$$\max f(w, r) \quad \text{and} \quad \sum f(w, r),$$

where w = waiting time and r = running time.

Waiting time refers to total time elapsed between submission and completion of a job. There are many possibilities for f, such as $f = w$ and $f = w/r$.

The first section of this chapter deals with optimization problems, using the result from elementary calculus that, except for boundary points and points without derivatives, $f' = 0$ at the extrema of f. The second section contains some models involving graphical optimization.

4.1. OPTIMIZATION BY DIFFERENTIATION

Maintaining Inventories

As a management consultant you are being asked for advice on production and warehousing policies. Where should you begin? One problem is the trade-off between storage space costs and setup costs for frequent small production line runs. In deciding how large an inventory of finished goods to maintain, a firm concerns itself with such things as cost of storage, setup expenses for a production run, discounts for bulk orders of raw materials, and orders lost as a result of lack of inventory. Because of the random nature of the time and size of orders, a probabilistic model is the most natural. We use a deterministic one, since the results are substantially the same if a firm receives many orders. For a fuller discussion of inventory problems, see R. L. Ackoff and M. W. Sasieni (1968), from which this model is adapted. See also the book by G. Hadley and T. M. Whitin (1963).

What should we optimize? We minimize the cost *per unit time* to the firm, subject to the constraint that all orders be filled. The only variable the manufacturer can control is the time between production runs. To begin with, we assume that the only costs the manufacturer adjusts by changing the production schedule are setup costs for production and storage costs for finished goods.

It is reasonable to assume that, when the production line is operating, it produces finished goods at a constant rate k per unit time. There is a cost c to set up the line at the beginning of a production run. This consists of profits lost by not using the production line for manufacturing at this time, various fixed costs, and any additional material and salaries that may be required. When the production line is not dedicated to the particular good we are interested in, we assume it can be used profitably for other work. We assume that the storage costs of the finished product are s per item per unit time, independent of the quantity stored. (This is reasonable if warehouse space can be used for other goods.) Finally, we approximate the discrete arrival of orders by a continuous arrival at a constant rate r per unit time. Discuss these assumptions and consider ways in which the model can be made more realistic. Remember that it is essential that the parameters in the model be determined if the model is to be of any use, and that this determination may be quite expensive for a complex model.

Let T be the length of time between one production run and the next. If t is the length of a production run, $kt = rT$; that is, goods produced equal goods sold during a cycle. Hence $t = rT/k$. If you graph inventory versus time from 0 to T, it rises from 0 to t with slope $k - r$ and falls from t to T with slope r. The area under the triangular curve is $A = (k - r)tT/2$ and is measured in units of items × time. Convince yourself that the storage cost is sA. Thus we want to minimize

$$(1) \qquad C = \frac{c + sA}{T} = \frac{c + s(k - r)tT/2}{T} = \frac{c}{T} + \frac{s(k - r)(rT/k)}{2}.$$

Differentiating with respect to T and setting the derivative equal to zero, we obtain $c/T^2 = s(k - r)r/2k$. From the form of (1) it is clear that C becomes infinite if T decreases to zero or increases to infinity, hence this extreme value of C is a minimum. Thus the optimum values for T and t are

$$T = \sqrt{\frac{2ck}{rs(k-r)}}, \qquad t = \sqrt{\frac{2cr}{ks(k-r)}}.$$

It is not obvious a priori that the optimal time varies as the square root of the setup cost and inversely as the square root of the storage cost per unit time.

We now consider storage costs for raw materials. Let's assume that there is only one raw material and that the precise amount needed is delivered at the beginning of the run. Let s' be the storage cost per unit time for enough raw material to produce one item of output. Convince yourself that the cost per unit time is

$$C = \frac{c + s(k-r)tT/2 + s'(rT)t/2}{T} = \frac{c}{T} + \frac{[s(k-r) + s'r](rT/k)}{2}.$$

Setting the derivative equal to zero, we obtain $c/T^2 = [s(k-r) + s'r]r/2k$. Thus the optimum values of T, t, and C are

$$T = \sqrt{\frac{2ck}{r[s(k-r) + s'r]}},$$

(2) $$t = \sqrt{\frac{2cr}{k[s(k-r) + s'r]}},$$

$$C = \sqrt{\frac{2cr[s(k-r) + s'r]}{k}}.$$

Since the model is only approximate and since we probably cannot determine the independent variables very accurately, it is important to have some idea of the cost incurred by making these errors. If T is replaced by αT, it is easy to show that the value of C is $(\alpha + \alpha^{-1})/2$ times the optimal value. For example, a 50% underestimate of T (i.e., $\alpha = 0.5$) increases C by 25%, while a 50% overestimate increases C by about only 8%—the same amount as a 33% underestimate would. We draw two conclusions from this. First, an error in choosing T does not change costs greatly unless the chosen value of T is quite far from the optimal value. Second, it is better to err on the high side than on the low side. Since the storage costs are the hardest to estimate and since T varies inversely with the storage costs [this follows from (2) and the fact that $k > r$], this suggests that underestimates of storage costs are better than overestimates. What we have done in this paragraph is an example of *sensitivity analysis*. Characterizing models as fragile or robust is a very crude form of sensitivity analysis.

We can use the results in (2) to determine how much warehouse space our company requires. (How is this done?) If this differs from the amount of space we now have, we should either get rid of excess space or acquire additional space. This is fine in the long run, but what do we do in the short run, that is, the period of time before we can change our warehouse space? Since the cost of the warehouse is fixed in the short run, s and s' should be zero. (See the discussion of the theory of the firm in Section 3.2 for an explanation of fixed and variable costs.) How can we determine the best short run plan? As pointed out at the beginning of this paragraph, if we knew the storage costs, we could use (2) to determine how much space is required. This suggests that we assign fake costs to make storage space needed equal to storage space available. The easiest way to do this is to replace s and s' by $s\sigma$ and $s'\sigma$, where σ is the factor that we have to scale costs by and s and s' are long run costs. (You should be able to show that this simply has the effect of multiplying the needed storage space computed from (2) by a factor of $\sigma^{-1/2}$.)

The situation with a bulk order of raw material is more complicated. Suppose a bulk order shipment consists of enough raw material to produce N finished items. For simplicity we assume that T is such that $p = N/rT$ is an integer; that is a raw material order lasts for p production cycles. (You may wish to study the model when p is not an integer.) The amount of raw material on hand is plotted in Figure 1 over p production cycles. The area under the curve is $N(pT - T + t)/2$. Combining this with $N = prT$ and $T - t = (k - r)T/k$, we see that the storage cost per unit time is

$$\frac{s'[N - (T - t)r]}{2} = s'\left(\frac{N}{2} - \frac{(k - r)Tr}{2k}\right).$$

Combining this with (1) we obtain the total cost per unit time:

$$C = \frac{c}{T} + \frac{rT(k - r)(s - s')}{2k} + \frac{s'N}{2}.$$

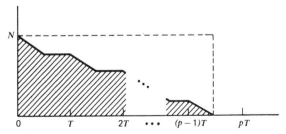

Figure 1   Raw material on hand during p production cycles of length T.

If $s \leq s'$, the best strategy is to make T as large as possible, that is, $p = 1$. When $s > s'$, we obtain the optimum values

(3)
$$T = \sqrt{\frac{2ck}{r(k-r)(s-s')}},$$

$$t = \sqrt{\frac{2cr}{k(k-r)(s-s')}},$$

$$C = \sqrt{\frac{2cr(k-r)(s-s')}{k}} + \frac{s'N}{2}.$$

This can be compared with the optimum nonbulk values given by (2) after a correction term is subtracted from the optimum bulk cost due to lower costs for raw materials. If the cost of materials is b lower per finished unit when the manufacturer orders in bulk, the correction term will be rb. Note that bulk ordering leads to longer productions runs, the ratio of times being

$$\sqrt{1 + \frac{s'k}{(s-s')(k-r)}}.$$

We have not discussed the possibility of allowing the warehouse to run out of finished goods and then back-ordering. This eliminates some storage costs at the expense of possibly losing some customer good will, hence some orders. Various approaches have been suggested. The following is adapted from B. L. Schwartz (1966). Most firms can expect to gain and lose customers at a fairly regular rate. At equilibrium the rate of loss and the rate of gain must be equal. What happens to these rates if the fraction f of delayed orders is increased? Since there will be more disgruntled customers, the rate of loss will increase. We assume that new customers are still gained at the same rate. This probably won't be true if f changes markedly, since bad reputations spread; however, it seems reasonable if f changes only slightly. The simplest model incorporating these ideas is

$$a(1 - f)N + bfN = constant,$$

where the constant is the rate at which new customers are gained, N is the number of customers, a is the probability of losing a customer whose order is filled promptly, and b is the probability of losing a customer whose order is delayed. Since r is proportional to N, it follows that $r(f)$ is proportional to $1/[a(1-f)+bf]$, and so

$$r(f) = \frac{r_0}{1 + f(b-a)/a}.$$

for some constant r_0. The storage costs must be reduced to reflect the fact that less storage space and time are used when $f \neq 0$. You should be able to show that $s(k - r)t/2$ is replaced by $s(k - r)(1 - f)^2 t/2$. This has the effect of replacing s by $s(1 - f)^2$ in the formulas obtained above for the optimum values of T, t, and C. Also, these values are now functions of f. When the selling price is p, the profit per unit time is $pr(f) - C(f)$. The optimum value of f can be determined by maximizing this function. Even in the simplest case this is quite messy. When the production line is so fast that we can approximate $(k - r)/k$ by 1, things are simplified a bit. Try it.

Geometry of Blood Vessels

The blood vessel system of higher animals is so extensive that evolution has probably optimized its structure. How much of the structure can we explain in this way? First, we need to know what is being minimized or maximized by optimization. We can say that the cost to the organism is minimized, but then we must say what we mean by "cost." This depends on the specific problem, so we'll put it off for the time being. Let's study the branching of vessels. For simplicity we consider only the case in which a vessel splits into two vessels, each of which carries equal amounts of blood. For the general situation of unequal-sized branches, see R. Rosen (1967, Ch. 3), from which this model is adapted.

Any reasonable model can be expected to lead to the conclusion that all three vessels lie in a plane, since otherwise we could shorten the lengths of all three simultaneously by making them planar—surely a saving for the animal. Structural considerations may prohibit this, but it is a reasonably accurate assumption, since sharp changes in direction are seldom required by structural constraints. By symmetry, the two smaller branches should have equal radii r' and flow rates f', and make the same angle θ with the larger vessel. Let r and $f = 2f'$ be the radius and flow rate of the larger vessel. See Figure 2. The organism has a "cost" associated with maintaining vessels and overcoming resistance in pumping blood. This cost per unit length is some function $C(r, f)$. Since we wish to minimize this, r and r' are determined as functions of f by

(4) $$\frac{\partial C(r, f)}{\partial r} = 0 \quad \text{and} \quad \frac{\partial C(r', f/2)}{\partial r'} = 0.$$

We also wish to choose θ to minimize the cost associated with the three vessels in the branch. If the vessels have lengths L, L', and L'', we wish to minimize

$$C = C(r, f)L + C(r', f')L' + C(r', f')L''.$$

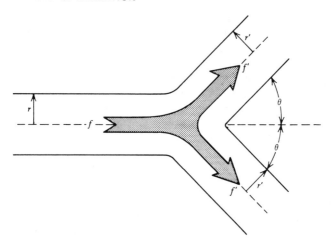

Figure 2   Arterial blood flow. Flow rates, f and f'; vessel radii, r and r'; branching angle, θ.

A slight change in L to $L + \Delta L$ results in a decrease in both L' and L'' by an amount equal to $\Delta L \cos \theta$ plus a term on the order of $(\Delta L)^2$. Draw a picture and convince yourself of this. Since $C' = 0$ at a minimum, ΔC must be on the order of $(\Delta L)^2$ or smaller. Hence

(5) $$C(r, f) - 2C(r', f') \cos \theta = 0$$

at an extreme point. This must be a minimum, since we can clearly increase the cost by increasing L so that θ approaches π^2. Since r and r' are determined by (4), this gives an expression for θ.

Let's consider a specific form for C. The work needed to overcome resistance in a rigid pipe with flow rate f and radius r is kf^2/r^4 per unit length, where k depends on the nature of the fluid. Vessel maintenance may depend on the space occupied by the vessel, the inner surface area of the vessel (where most of the wear may occur), the volume of the cells making up the vessel, or some combination of these. The first two give a cost per unit length proportional to r^2 and r, respectively. The third depends on how the thickness of the vessel wall varies with r. If it is proportional to r, the cost per unit length is proportional to r^2. If it is constant, the cost per unit length is proportional to r. In order to include all these possibilities for vessel maintenance in some simple fashion, we consider a contribution of the form Kr^a, where $1 \leq a \leq 2$. The total cost per unit length is thus $kf^2/r^4 + Kr^a$. By (4) we have $f^2/r^{a+4} = \kappa$, where $\kappa = aK/4k$. Thus $C(r, f) = \lambda r^a$ and, since $(f/f')^2 = 4$,

(6)
$$\left(\frac{r}{r'}\right)^{a+4} = 4.$$

Equation 5 yields

$$\cos \theta = \frac{(r/r')^a}{2} = 2^{(a-4)/(a+4)}.$$

Since $2 \geq a \geq 1$, it follows that $37° \leq \theta \leq 49°$. As far as I know, this has not been tested. However, it is known not to hold at the capillary level. If you are interested in obtaining some data, the illustrations in F. H. Netter (various dates) could be measured. I've been told that his drawings are quite accurate.

By using (6) plus the known radii of the aorta and capillaries we can determine the number of branchings between a capillary and the aorta in an organism: If there are n branchings, by (6) the ratio of the aortic radius to the capillary radius equals $4^{n/(a+4)}$. Rosen gives an approximate value of $10^3 \approx 4^5$ for this ratio in dogs. Hence $n \approx 5(a + 4)$, which ranges from 25 to 30. Since the number of capillaries equals 2^n, there are between 3×10^7 and 10^9 capillaries. An empirical estimate cited by Rosen is 10^9.

Fighting Forest Fires

Your state forestry service wants to reduce the financial and environmental costs of forest fires. How can they do this? What is the best way to reduce the cost of forest fires within the limits of present fire control methods? The following is an adaptation of a model presented by G. M. Parks (1964) for determining the size of an optimal fire fighting force. Another possibility that needs serious consideration is increasing the effort spent on detection; however, we ignore it here. "The best way" is interpreted to mean the least costly way. This means we must assign costs for the burned area and the injuries and deaths of fire fighters. The first cost is very difficult to assess; outdoorsmen, lumbermen, and city dwellers are likely to assign quite different costs. In California in 1963 "current practice [assigned] ... values from $25 to upwards of $2,000 per acre." What about the second cost? Since more fire fighters mean quicker control of a fire, there is less chance per fighter for injury; furthermore, fire fighters are assumed to receive monetary compensation. Therefore we do not consider the cost of injuries and deaths.

Let $B(t)$ be the area burnt by time t, where time is measured from $t = 0$ at time of detection. We assume that the fire has stopped when $B'(t) = 0$.

Let T_a be the time the fire is first attacked and T_c the time it is brought under control. Thus T_c is the least $t > 0$ such that $B'(t) = 0$. Let x be the size of the fire fighting force (assumed constant from T_a to T_c). The costs for fighting a particular fire are:

C_b, the cost per acre of fire (burnt acreage plus cleanup expenses).
C_x, the cost in support and salary per fire fighter per unit time.
C_s, one-shot costs per fire fighter (such as transportation to and from the site).
C_t, costs per unit time, while the fire is burning, for maintaining the organization on an emergency basis, redirecting traffic, and so on.

(Note that we are implicitly assuming that all the C are constants.) The total cost is

$$C = C_b B(T_c) + (xC_x + C_t)(T_c - T_a) + xC_s.$$

To minimize C as a function of x, we must determine $B(t)$. We assume that each fire fighter reduces the burning rate of the fire at a constant rate E, that is, decreases $B''(t)$ by E. Thus

(7a) $$B'(t) = b(t), \quad \text{for } t < T_a,$$

(7b) $$B'(t) = b(t) - E(t - T_a)x, \quad \text{for } T_a \le t \le T_c,$$

where $b(t)$ is to be determined. Parks simply assumes that $b(t)$ is a linear function of t. We can derive this from the crude assumption that the fire is spreading circularly at a uniform rate: The perimeter is proportional to $b(t)$ and the rate of change of the perimeter is a constant. Thus $b(t) = G + Ht$.
    Criticize the model.

    To find T_c we set $B'(t) = 0$ in (7b) and obtain

$$T_c = T_a + \frac{G + HT_a}{Ex - H}.$$

Note that $Ex > H$ is required if the fire is ever going to be stopped. We now integrate (7) to obtain

$$B(T_a) = B(0) + GT_a + \frac{HT_a^2}{2}$$

$$B(T_c) = B(T_a) + \frac{(G + HT_a)^2}{2(Ex - H)}.$$

For convenience let $b_a = b(T_a) = G + HT_a$ and $z = x - H/E$, the number of fire fighters above the bare minimum. By combining the above results we obtain

$$(8) \qquad C = C_0 + C_s z + \frac{[(HC_x/E) + C_t + (C_b b_a/2)]b_a}{Ez},$$

where C_0 is a constant. Setting the derivative with respect to z equal to zero, we obtain the optimal value:

$$(9) \qquad x^* = b_a \sqrt{\frac{C_b + 2C_t/b_a + 2HC_x/Eb_a}{2C_s E}} + \frac{H}{E}.$$

The values of C_x, C_s, and C_t can be determined for a region; the values of C_b can be tabulated for various types of forests; the values of H and E can be tabulated by type of forest and wind conditions; and b_a can be determined on the spot. Then (9) can be applied. It is unlikely that this would be done by the forest service; however, (9) could be used to make general recommendations to forestry officials. Parks has done this. He obtained numerical estimates and concluded that 102 of the 139 fires in the Plumas National Forest in California in 1959 were undermanned. In particular, the model predicts that the four fires that burned over 300 acres each would have burned less than 100 acres each with proper manning.

There are problems with relying on (9), even if we believe that the model is correct and are able to reach some agreement on estimates for the various costs. It is still necessary to know b_a, H, and E. Unfortunately b_a tends to be underestimated because that makes the lookout appear more alert, while H and E are dependent on so many factors that good estimates in a particular situation may be hard to obtain even if tables are prepared ahead of time. How sensitive are (8) and (9) to such errors?

The graph of x versus C in Figure 3 shows that underestimating x^* by a large amount is more expensive than overestimating it. The critical variables are H and E, since errors here can shift us into the untenable position of fielding less than H/E fire fighters. We could improve the situation somewhat by tabulating H/E instead of H and E separately. (Of course we also need either H or E as well, but this way we are spared the necessity of dividing two uncertain quantities to obtain the critical quantity H/E.) What *is* H/E? It is the number of fire fighters needed to keep a fire from spreading at a faster rate, that is, enough fire fighters so that $b(t)$ is a constant. Not only does this sound hard to measure, it sounds impossible. Surely the number of fire fighters must depend on the size of the fire. According to the model the number of such fire fighters is independent of the size of the fire. Before we

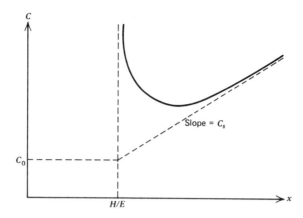

Figure 3 Firefighting cost as a function of manpower.

accept the model it would be a good idea to check this counterintuitive prediction, since H/E plays such a crucial role in determining x^*. As far as I know, this hasn't even been noted, much less explored.

PROBLEMS

1. Returning to the blood vessel model developed above, do you think Rosen's data on the number of capillaries is strong evidence for the cost function $C = \lambda r^2$? Why? Propose further tests for the theory that evolutionary pressure has led to minimal total cost and that the cost per unit length is $C(r, f) = \lambda r^a$, with $1 \le a \le 2$. How can a be estimated?

2. Suppose you wish to get from one place to another in the rain by traveling in a straight line. How fast should you walk (or run) to stay as dry as possible? The following model is due to B. L. Schwartz and M. A. B. Deakin (1973).

   (a) Let's approximate a person by a rectangular prism (a box) with a ratio of areas given by

   $$\text{front:side:top} \quad :: \quad 1:\eta:\varepsilon$$

   Assume that the rain's velocity is $(w, W, -1)$ and that the person's is $(v, 0, 0)$, where the z coordinate is vertical upward. Show that the amount of rain hitting the person per unit time is proportional to $|w - v| + \varphi$, where $\varphi = |W|\eta + \varepsilon$, a constant.

(b) Show that, if $\varphi > w$, you should·run as fast as possible and that otherwise you should run with $v = w$ or as fast as possible, whichever is slower. This has a simple interpretation in terms of keeping your front and back dry. What is it?

(c) Criticize the model. Can you improve it? How do the new and old predictions compare?

3. Suppose you are an advisor to a congressperson who wishes to develop legislation to regulate commercial fishing so that the fish populations will be preserved. To advise him or her you need to become familiar with the economic aspects of the problem. This material is adapted from C. W. Clark (1973). Let N be the size of the fish population. For simplicity, assume a selling price of p per fish, independent of the quantity sold.

(a) Argue that the harvest cost per fish $c(N)$ is a decreasing function of N and that, if there are no fishing regulations, we can expect the fish population to be at the level N_f, where $p = c(N_f)$. Cost includes salaries, fuel, income lost because capital is tied up in boats, and so on.

(b) Suppose we assume a simple reproduction model: $N' = g(N)$. Show that a reasonable shape for g is a concave arc passing through $N = 0$ and $N = N^*$, the maximum population that can maintain itself when there is no fishing. Show that maximum sustained yield is obtained at N_m, the solution of $g'(N_m) = 0$. What is the yield? What does $N^* \leq N_f$ say about the economic feasibility of fishing? What about $N^* \geq N_f$?

(c) Suppose the fish population is to be maintained at the most profitable level. Call this N_p. Show that profits are given by

$$P(N) = g(N)[p - c(N)],$$

and that N_p is the solution of $P'(N_p) = 0$.

(d) What can you say about the relative sizes of N_f, N^*, N_m, and N_p?

(e) Under what conditions is it economically best to drive the species to extinction by fishing? *Hint*: Perhaps the left hand zero of $g(N)$ should be at a point to the right of zero, since a dispersed population below a certain critical level may not be able to come together to reproduce. If extinction is not economically feasible, is legislation a good idea anyway? Explain. What about fishing in international waters, for example, whaling?

(f) Can you improve the model? What if p depends on harvest size?

(g) Apply the above ideas to buffalo hunting (previous century), deer hunting (present day), tree farming, and anything else you'd care to.

   Notes: A graphical approach to parts of the above problem may be helpful. See Chapter 3 and Section 4.2. Fisheries have been studied extensively. Among the journals devoted to the subject are *Fishery Bulletin* and *Transactions of the American Fisheries Society*. See also C. W. Clark (1976).

4. In designing a multistage rocket, how would you decide on the number and size of the various stages? By having multiple stages, unneeded fuel containers can be discarded, thus reducing the amount of mass that must be accelerated for the rest of the flight. Unfortunately there is a cost: Additional motors are needed so that each stage will have an engine, and this adds to the weight until the motor is discarded. Clearly some compromise should provide the biggest payload (or longest flight) for the money. For simplicity we assume that cost is proportional to weight. Therefore we maximize the terminal velocity for a given initial mass and a given payload mass. Again for simplicity let us neglect the effect of gravity. (The crude assumptions we are making can be removed, but then the optimization problem may require a computer.) We need the physical fact that the mass m and the velocity v of a rocket with constant exhaust velocity v_e are related at any time by

$$m \exp\left(\frac{v}{v_e}\right) = \text{constant},$$

when gravity and air resistance are neglected. The constant changes each time the rocket drops a stage. (For those who wish to derive the result, it is simply a conservation-of-momentum argument: $m\,\Delta v + v_e\,\Delta m = 0$.) To begin with, let's find the optimal division between stages, given that we are to use n stages and the payload counts as a stage. Let

M_i be the mass of the entire rocket (including fuel) when the ith stage begins to fire.
F_i be the mass of the fuel in the ith stage.
C_i be the mass of the fuel casing in the ith stage.
R_i be the mass of the rocket motor and other support in the ith stage.

By assumption, M_1 and M_n are given, $F_n = C_n = 0$, and we can assume that $R_n = 0$ by absorbing it in the payload.

(a) Show that the terminal velocity is

$$v_T = v_e \sum \log\left(\frac{M_i}{M_i - F_i}\right).$$

(b)   Using (a), show that, if M_j is such that for given values of M_{j+1} and M_{j-1},

$$\log\left(\frac{M_{j-1}}{M_{j-1} - F_{j-1}}\right) + \log\left(\frac{M_j}{M_j - F_j}\right)$$

is a maximum and, if this holds for $2 \leq j \leq n - 1$, the rocket maximizes v_T. *Remark*: This uses an important idea in maximization: A solution that is locally a maximum is often globally a maximum. In this instance, if the division of mass $M_{j-1} - M_{j+1}$ between stages $j - 1$ and j is the best possible for all j, the entire rocket is the best possible.

(c)   We assume that $C_i \propto F_i$ and $R_i \propto M_i$, with constants of proportionality independent of i for $1 \leq i \leq n - 1$. Discuss. Use this to conclude that $F_i = aM_i - bM_{i+1}$ for some a and b and thus express $\log\left[M_i/(M_i - F_i)\right]$ in terms of M_i and M_{i+1}.

(d)   Using (c), reduce the expression in (b) to a function of the single variable M_j. Show that it is a maximum when

$$\frac{M_j}{M_j - F_j} = \frac{M_{j-1}}{M_{j-1} - F_{j-1}}.$$

Conclude that v_T is a maximum when $M_j/(M_j - F_j)$ is constant for $1 \leq j \leq n - 1$. Interpret in terms of Δv for each stage.

(e)   How can you determine the optimum value for n, the number of stages? How does the reliability change as the number of stages increases? What can you do about this and how does it affect the model?

(f)   Can you propose a more realistic model which can be analyzed easily?

(g)   What additional factors would you take into account if you were actually attempting to design a multistage rocket?

5.   A troubleshooter spends a lot of time flying in his private plane to various industrial plants which he helps out. He wishes to spend the least amount of time possible traveling. Where should he live? Of course, you need data. What data do you need? You should set up a model so that data collection is feasible. How would you change your approach if he used commercial airlines?

6.   What is the best strategy for a swimming fish to adopt if it wishes to travel with the least expenditure of energy? (This "wish" is not conscious, but rather a result of natural selection.) Since the motions involved in swimming increase the drag on a fish to about three times its value when

the fish is gliding, it is to the fish's advantage to keep swimming time down. This leads to burst swimming (D. Weihs, 1973, 1974). Fish that are heavier than water can alternate between swimming upward and gliding downward. We study the simplest case of this discussed by D. Weihs (1973).

We assume that the fish attempts to move with a constant velocity v. (Other assumptions are possible, but this seems fairly reasonable, and we can handle it.) Let D be the drag on the gliding fish at this velocity and kD the drag on the swimming fish. Let W be the net weight of the fish in water, α the angle of downward glide, and β the angle of upward swimming. Thus we're assuming that the fish travels along a path which, when viewed from the side, has a sawtooth appearance. We assume that the energy used by the fish per unit time above and beyond that required simply to stay alive is proportional to the force it exerts in moving.

(a) Criticize the assumptions.

(b) Show that $W \sin \alpha = D$ and that the swimming force is $kD + W \sin \beta$.

(c) Show that the ratio of energy in the burst mode to energy for continuous horizontal swimming to go from a point A to another point B is

$$\frac{k \sin \alpha + \sin \beta}{k \sin (\alpha + \beta)}.$$

(d) It has been found empirically than $\tan \alpha \approx 0.2$. What is the best value for β? How much energy does the fish save? How important is it that the fish estimate β accurately? (We should answer this because it may be unrealistic to expect accurate estimates.)

(e) Criticize the model.

(f) Suppose the fish wishes to swim from A to B in a given time. Construct a model. Drag is roughly proportional to v^2. The energy per unit time (power) used to overcome drag in swimming is nearly proportional to v^3.

7. Two firms Y and Z are competing for a market. If Y spends y per unit time on advertising and Z spends z, we could expect that Y's share of the market in the long run is a function of the total advertising attributable to Y; that is, $f[y/(y + z)]$ for some function f. If the two firms are similar, Z's share of the market will be $f[z/(y + z)]$.

(a) Criticize the above suggestion.

(b) Show that, for $0 \le x \le 1$, $f(x) + f(1 - x) = 1$ and $f'(x) = f'(1 - x)$.

(c) Assuming the above, how should Y and Z act so as to maximize profit—assuming there is neither tacit nor explicit collusion between the two firms. How reliable is the prediction? You can assume that all costs and the function f are known.

This problem was adapted from R. G. Murdick (1970, Ch. 2).

8. What is the optimum number of years a company should keep trucks in its fleet before buying new ones? This can lead to many complications as the model becomes more and more realistic. Begin with a very simple model in which the main factor is rising maintenance costs. You can work up to as complicated a model as you feel the situation warrants.

4.2. GRAPHICAL METHODS

For the reasons given in Section 3.1, this section is limited to *qualitative* problems with few variables. The idea is simple: We wish to maximize a function f like "fitness" or "happiness," subject to certain constraints. The constraints and the curve f = constant are plotted, and the point where f is maximized is read from the graph. When the problem can be stated in clear, quantitative terms, more sophisticated methods such as Lagrange multipliers and mathematical programming are used.

A Bartering Model

Suppose two people have two goods which they wish to use in bartering with each other. What can we say about the situation? We assume there is some satisfaction associated with various mixes of the goods, and each person wishes his or her satisfaction to be as great as possible. For example, if I have 25 inches of French bread and you have 20 ounces of wine and it is lunch time, we will probably be able to work out a trade in which both of us will be better off. (Don't suggest simply "sharing"—that's frowned upon in *economic* models.) Can we say more about this? Let's consider another situation. Suppose I have 2 yards of one fabric and you have 2 yards of another. We may not wish to do any trading unless we switch ownership completely, because anything else would lead to rather small pieces of fabric. Can a model explain *both* situations?

We begin with the concept of *indifference curves*. I may say that as far as I am concerned 10 inches of bread and 4 ounces of wine together are just as good as 6 inches and 10 ounces. We say that I'm indifferent between (10, 4) and (6, 10). The set of points that I consider to be indifferent to (10, 4) form a set which is usually a curve. It is called an *indifference curve*. Several of my indifference curves are sketched in Figure 4. Can you explain the shape? A curve further toward the upper right contains points of greater

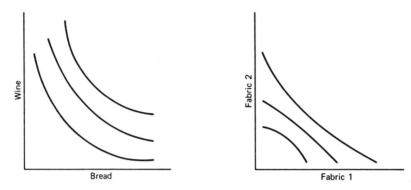

Figure 4 Two types of indifference curves.

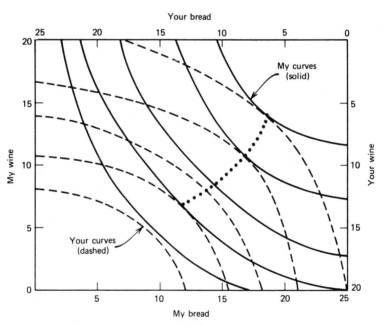

Figure 5 An Edgeworth box: our joint indifference curves. Dotted line is bargaining path.

satisfaction to me. (Why?) Thus I want our bartering to lead to a point on a curve far toward the upper right.

Now let's put your indifference curves and my indifference curves together. I've done this in Figure 5 for bread and wine. Note carefully the labeling of the axes: Altogether there are 25 inches of bread and 20 ounces of wine, and any point in the rectangle describes some division of the bread and wine between the two of us. Now suppose I agree to stay on one of my indifference curves. How can you maximize your satisfaction? The answer is simple: Choose a point where one of your indifference curves is tangent to mine. Another way of viewing this is to say that, if our indifference curves are not tangent at the point we have selected, there is another point where neither of us is worse off and at least one of us is better off. Hence we should stay on points of tangency. This is the *bargaining path*, which is shown dotted in Figure 5. It starts on my indifference curve containing (25, 0) and yours containing (0, 20), because neither of us will agree to be worse off after trading. Where on the curve we end up depends on our bargaining abilities. (Various people have attempted to be more specific.) Figure 5 is called an *Edgeworth box*.

What about the yard goods case? Here the indifference curves have a different shape, so that the points of tangency give minima instead of maxima. Thus we do better at the boundary.

What if we are trading more than two goods? For three goods we can still picture the situation: There are *indifference surfaces*, but the points of tangency still form a curve. This is true for any number of goods. We can put this result in a somewhat surprising form:

Suppose Bill and Mary are trading and I know their preferences. If Bill tells me how much of one of the goods he has settled for, I can then say, "Unless you have settled for the following amounts of the remaining goods, you and Mary can arrange a trade that would be better for both of you."

This model has several drawbacks. First, to make it quantitative requires a great amount of experimental work gathering data; however, psychologists have collected data of this sort in past experiments. Second, the indifference curves may shift with time—I may be more interested in wine after haggling with you for a while. Third, I may derive satisfaction from how well or how poorly I feel you are doing. Can you think of other objections? Do you think these ideas on bargaining would be useful in bilateral trade negotiations between the United States and Japan? In arms limitation talks between the United States and the U.S.S.R.? Discuss your reasons.

Changing Environments and Optimal Phenotype

Why do some animals have only a few quite distinct forms for different situations (e.g., queen, drone, and worker forms among honeybees), while others exhibit a whole range of variation (e.g., variation in the size of many plants with the climate)?

Suppose a habitat consists of two distinct types of environments. Examples are: oak trees versus maple trees (relevant for plant eating insects): warm versus cool weeks (relevant for insects producing more than one generation per year); and the nest versus the outdoors (relevant for some social insects with castes like ants). Assume that the animal or plant spends most of its life in only one of the two environments and that for developmental or genetic reasons the organism can end up having one of several *phenotypes*. We want a model that explains why some organisms have markedly different phenotypes in different environments while others do not. The following ideas are adapted from R. Levins (1968, Ch. 2). See E. O. Wilson and W. H. Bossert (1971, pp. 73–77) for related material.

We begin with the idea of *fitness*. In vague terms, the fitness of an individual is a measure w of its expected success. This could be measured in terms of the extent to which an individual's genes survive and spread in future generations or, for social insects with a single queen, the survival and reproduction of the nest. Thus fitness could be measured by the expected number of descendants at some future time. Even this is rather vague, because fitness is a very slippery concept to try to grasp precisely. We can allow it to remain vague as long as we are aware that we are doing so, because we only wish to make crude qualitative predictions. Since we can't obtain the data that would be required by a quantitative model anyway, it is pointless to attempt to formulate such a model. *The essential property we demand is that the fitness down to the nth generation is the product of the fitnesses at each generation.*

Suppose the fitness of an individual in the first environment is W_1, and in the second W_2. If the fraction of time spent in the first environment is p, the fitness after n generations is

$$(10) \qquad W_1^{pn} W_2^{(1-p)n} = (W_1^p W_2^{(1-p)})^n.$$

We wish to maximize (10), subject to the constraint that the fitnesses W_1 and W_2 are actually possible. The shaded regions in Figure 6 indicate fitnesses of biologically possible individuals. The regions are called *fitness sets*. On the left, the environments are sufficiently similar so that an intermediate individual A can do well in both. In contrast, the intermediate individual B on the right does poorly because the environments are too dissimilar.

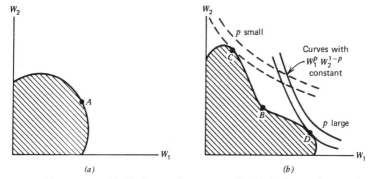

Figure 6 Fitness sets. (*a*) Similar environments. (*b*) Dissimilar environments. Intermediate individuals such as *A* and *B* occur only in the case of similar environments.

To maximize (10) we simply plot curves on which $W_1^p W_2^{1-p}$ is constant and note that the optimum individual occurs at the point where such a curve is tangent to the fitness set. The curve has a shape similar to that of the hyperbola $xy = c$.

As p varies, the curves on which (10) is constant vary in shape. When the two environments are similar, the optimum varies smoothly with p. In dissimilar environments, there may be a sudden jump from the specialist C (in Figure 6*b*) to the specialist D as p increases, completely avoiding the poor generalist B. Examples of both situations occur. You should be able to think of many examples of the former, for example, variation in thickness of coat in furbearing animals with climate. Here's an example of the latter: Some species of butterflies mimic other species that are distasteful to predators. There is a species in South America that mimics different species in different parts of its range. An organism with the phenotype of a compromise mimic would be poorly protected.

Let's consider caste formation in ants. The first environment is the nest defense milieu, and the second is the nest maintenance milieu. In Figure 7 is plotted a soldier (*S*), a worker (*W*), and two possible generalists. If defense and maintenance were sufficiently different so that *G* is the best possible generalist, there would be evolutionary pressure toward caste formation. If *G'* were possible, castes would be unlikely to form. If defense were rare, evolution might lead to the castes *G'* and *W*.

Note that we haven't discussed the shape of the fitness curves in connection with Figure 7. It's rather tricky; in fact, this whole subject is a bit tricky. You may want to work on it.

E. O. Wilson (1975, pp. 306–309) presents another approach to caste formation which we discuss briefly. It involves some simple probability

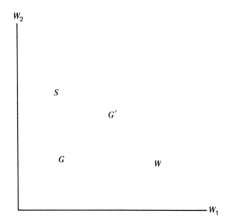

Figure 7   Caste formation. Soldier, S; worker, W; generalists, G and G'.

theory. Suppose we have a list of the possible castes and a list of the situations (e.g., repel an attack, forage) that a colony must deal with. A colony cannot fail too often and still survive. Various castes contribute more to success in a particular situation than others do. Let P_{ij} be the probability that caste i will fail to deal with problem j. We assume that the castes contribute independently to success, so that $P_{1j}P_{2j}\dots$ is the probability that problem j will not be dealt with successfully by the colony. One way to limit failures is to require that

$$(11) \qquad \prod_i P_{ij} \le M_j, \qquad \text{for all } j.$$

Clearly P_{ij} depends on the number of members in caste i. The simplest assumption is, again, independence:

$$(12) \qquad P_{ij} = p_{ij}^{n_i},$$

where n_i is the number of individuals in caste i. If c_i is the cost of producing and maintaining a member of caste i averaged over the individual's lifetime, we can describe the colony's problem as follows:

$$(13a) \qquad \text{Minimize:} \sum_i c_i n_i,$$

$$(13b) \qquad \text{Subject to:} \, n_i \ge 0$$

$$(13c) \qquad \text{And:} \sum_i n_i \log p_{ij} \le \log M_j.$$

[The last expression comes from combining (11) and (12).] This is an example of a problem in *linear programming*, a field in which a variety of textbooks exist.

This model has several drawbacks. The major ones are probably the (highly unrealistic?) assumptions of independence leading to (11) and (12). Also, the constraints in (11) may not be an appropriate way to define not failing too often. Some of the difficulties can perhaps be avoided by redefining terms. Others require revisions that would destroy the linearity of (13c). Can you suggest ways to improve the model?

Let's illustrate the model by considering a simple case involving only two possible castes. Introduce two axes which indicate the number of members in each caste. Constraints (13b) limit us to the first quadrant. Each of the constraints (13c) requires that we look above a line of slope $-\log p_{i1}/\log p_{i2}$ and a given intercept. Figure 8 illustrates a possible configuration with four problems. Since $c_1 n_1 + c_2 n_2$ is constant on straight lines of slope $-c_1/c_2$, picking out the point in the shaded region that produces a minimum is fairly easy to do graphically. You should be able to describe a method. Note that it is possible to obtain a solution in which not all castes actually exist; that is, $n_i = 0$ for some i. This is as it should be.

While these models are still quite crude, there is hope that this approach may shed light on why some species of social insects have more castes than others and why the energy of a colony is divided between castes in the way

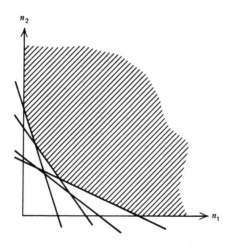

Figure 8   A linear caste formation model. Inequalities (13b) and (13c) hold in the shaded region.

that it is. (The *why* of sociality in insects is an interesting question which is beginning to be answered. See E. O. Wilson (1975, pp. 415–418) for a discussion.)

PROBLEMS

1. Two college administrators are trying to decide on an admissions policy so as to obtain the "best" possible students for their college. They each have different ideas on how important various traits are in a good student. Can you suggest a theoretical plan for helping them? A practical one? What if three administrators are involved? *Note*: The time and money required for extensive testing are not available; only the administrators and their opinions are available.

2. Let's consider a bread and wine problem different from the one in the text.

    (a) Suppose I am buying lunch, wine costs 10 cents per ounce, bread costs 5 cents per inch, and I have \$1 to spend. If I know my indifference curves, how can I determine what to buy?

    (b) Suppose the price of wine rises to 13 cents. What will happen to the amount of wine I buy? The amount of bread?

3. How do wages affect the amount of time a person works? An individual wants both income and leisure time. Hence he or she is willing, up to a point, to work longer when the hourly wage is higher. As the wage becomes higher, however, an income saturation effect occurs and the worker may wish to work somewhat less time as the wage rate increases, thereby increasing both leisure time and income. A reverse effect may occur if the wage is low, since a person often desires a certain level of income and may more readily sacrifice leisure to attain it if wages are raised slightly. Can we cut through this complexity to decide if, as an employer, it is better for you financially to offer overtime or higher wages?

    (a) Using the coordinates hours per day and dollars per day, plot indifference curves for a worker. What is the shape of such a curve? *Hint*: What does the slope mean?

    (b) For a particular hourly wage rate a straight line through the origin gives hours worked versus wages received. Why? Describe geometrically how to measure the number of hours a worker would choose to work if he or she were given the freedom to choose (e.g., a self-employed person such as a lawyer or a plumber.) As the hourly rate varies, the optimum point varies. Describe and interpret the locus.

(c) Discuss the effect of overtime.

(d) Is is better from the employer's viewpoint (i.e., maximum number of hours per employee for a given total wage) to raise wages or to raise overtime pay? Why?

(e) Instead of considering a single worker, carry out the above analysis for the entire work force potentially available to the employer.

For further discussion of this topic see K. J. Cohen and R. M. Cyert (1965, Ch. 5).

4. Suppose you are faced with the problem of how to adjust traffic signals for rush hour traffic. What is the best way to do it? This problem is adapted from D. C. Gazis and R. B. Potts (1965). We suppose that at $t = 0$ there is no line at the signal in either direction. At the end of the problem, try to decide how important and how realistic this assumption is. Cars arrive at the signals at rates $q_N(t)$ from the north and $q_E(t)$ from the east. The signal can handle cars at a rate k. Let $Q_N(t)$ and $Q_E(t)$ be the integrals of q_N and q_E from 0 to t.

(a) Show that T, the earliest possible time the intersection can be cleared, is determined by the equation

$$Q_N(T) + Q_E(T) = kT.$$

(b) Let $f_N(t)$ be the flow of the north cars through the intersection at time t. Define F_N, f_E, and F_E in the obvious fashion. What relationships can you discover among the four functions just defined? Interpret the area between the curves Q_N and F_N in terms of delay time.

(c) Show that the total delay time at the intersection is a minimum if and only if both intersections are cleared simultaneously at time T. Determine T.

(d) What is the best form for F_N? Suppose the rush hour traffic starts to arrive earlier from the north so that $q_N(t)$ is large when t is small but $q_E(t)$ is small when t is small. Consider other situations, too.

(e) Discuss improvements and generalizations for the model. You need not limit yourself to graphical methods. Among the problems you could consider are flows from all four directions, lost time when signals change, unequal rates of flow (the parameter k) in different directions.

5. In industrial chemical processes, yield is frequently highly dependent on temperature and pressure, but these are limited in range by technological and economic considerations. The amount of impurities also depends on temperature and pressure. Describe a graphical approach for

obtaining maximum yield when one impurity cannot exceed a certain value. Do the same for several impurities. This idea is discussed in B. Noble (1971).

6. (a) Consider the following model of political behavior. There are three voters, two issues, and two politicians. Suppose the positions taken on the issues can be represented as points on a plane and that the indifference curves of each voter are circles centered about the, to him, ideal position. Show that the politician who declares his positions last can ensure himself at least two of the three votes.

   (b) Can you construct a more realistic model? What are its political implications? How much faith do you have in the predictions? Why? See R. D. McKelvey (1973) for further discussion.

CHAPTER 5

BASIC PROBABILITY

Most of the models in this book are deterministic. Stochastic models are discussed here and in Chapter 10. Here we use only basic discrete probabilistic concepts, but more sophisticated concepts, such as the central limit theorem, are needed in Chapter 10. The Appendix contains a terse discussion of the probabilistic concepts required. It can serve as a refresher or as a reference for a more leisurely classroom discussion.

5.1. ANALYTICAL MODELS

Sex Preference and Sex Ratio

Some people have expressed concern about the possibility of a population markedly altering its sex ratio (number of males divided by number of females) because of preferences for children of a particular sex. This could be a real problem if intrauterine sex determination is coupled with abortion or if infanticide is practiced. To what extent can a population affect the sex ratio purely by means of birth control, including abortion which is not related to the sex of the fetus? The following discussion is based on L. A. Goodman (1961).

Let's ignore multiple births to make the analysis easier. They are sufficiently rare that the effect on the model will be quite small.

We must say something about the chances that a healthy baby born to a given couple will be a girl or a boy. This may vary from couple to couple. One can give a reasonable biological argument that it does not depend on the sex of the children already born to the couple. There are data indicating that sex is slightly related to the age of the couple. Since this is not easily incorporated in a model and since it has only a slight effect, we ignore it.

The major problem is: How many children is a couple able to bear? This is a thorny problem. We ignore it completely in the following discussion and consider it briefly in Problem 1.

Our assumptions can be summarized as follows:

1. There exists a probability p_i that a child born to the ith couple will be male and a probability $q_i = 1 - p_i$ that it will be female. The value of p_i is not a function of the sex of the other children of the couple and cannot be adjusted by the couple.
2. Each birth leads to exactly one child.
3. A couple can have as many children as desired.

In view of assumption 3, a couple can have additional children if a child should die any time after it is born. Hence we can ignore deaths in childhood and interpret p_i as being the probability that a child who is born *and survives through childhood* is a male. After reading the following discussion, comment on the realism of the assumptions and try to determine what effect they have on the conclusions. In particular, Problem 1 asks for a discussion of a model in which assumption 3 is replaced by an upper bound on the number of children per couple. In technical terms, the model proposed treats sexes of children born to a couple as Bernoulli trials.

We wish to study the value of μ, the fraction of males in the population. Let F_i be a random variable equal to the number of females born to the ith couple, let M_i be the number of males, and set $N_i = F_i + M_i$. Then

$$(1) \qquad \mu = E \frac{\sum M_i}{\sum N_i} \approx \frac{\sum E(M_i)}{\sum E(N_i)},$$

where E denotes expectation. Approximating the expectation of the ratios by the ratio of the expectations, as was done in (1), is quite accurate for large populations. (If you have had a course in mathematical probability theory, you might like to prove it.)

In view of assumption 1, the expected fraction of boys born to the ith couple will be p_i. Hence $E(M_i) = p_i E(N_i)$, and *there is no way a couple can change the expected fraction of boys born to it.* From (1) we have

$$(2) \qquad \mu \approx \frac{\sum p_i E(N_i)}{\sum E(N_i)}.$$

It follows from (2) that the population can cause a change in μ only by introducing a correlation between p_i and $E(N_i)$. When there is no sex preference, it is reasonable to assume that $E(N_i)$ and p_i are uncorrelated. In this case the right side of (2) equals the average of p_i over all couples.

What values of μ are possible? Since (2) is a weighted average of the p_i, the value of μ must lie between min p_i and max p_i. Because of assumption 3, the population *working as a whole* can approach any value within these limits. Also, the population working *individually* can *approximate* any λ between min p_i and max p_i: A couple continues to have children as long as the fraction of males in the n children they already have does not differ from λ by more than $n^{-1/3}$. In general, the closer some p_i are to λ, the closer we can expect μ to approximate λ. The choice of $n^{-1/3}$ is somewhat arbitrary. We want a function that tends to encourage couples with p_i close to λ to have many children. Since the fraction of children that are males tends to differ by something on the order of $n^{-1/2}$ for random reasons, we want a function that is large compared to $n^{-1/2}$ for large values of n. The function $n^{-1/3}$ is such a function.

Using (2) we argued that min $p_i \leq \mu \leq$ max p_i. There is an error in this argument: (1) is an approximation that is accurate only for large populations, and so only the approximation to μ lies between min p_i and max p_i. To see that μ need not lie within these limits, consider a population consisting of a single couple using the rule, "Stop after one child if the first child is a boy, otherwise have two children." Set $p_1 = p$ and $1 - p = q$. The possible sequences of children are M, FM, and FF, and their probabilities are p, qp, and q^2, respectively. Thus

$$\mu = 1p + \tfrac{1}{2}qp + 0q^2 = p + \tfrac{1}{2}pq > p.$$

When $p = \frac{1}{2}$, this equals 0.625. Now suppose there are k couples all using the same rule and all with $p = \frac{1}{2}$. The expected sex ratio for $k = 1, 2, 3, 4, 5$ is 0.625 (as just computed), 0.563, 0.541, 0.530, and 0.524, respectively. Thus the approach to $\frac{1}{2}$ is fairly slow.

The above discussion shows what *can* be achieved as values for μ. What *will* be achieved if each couple independently pursues a plan based on a desire for children of a given sex? Many plans are possible. Three examples are

1. A couple may continue to bear children until they have a child of the desired sex.
2. They may continue bearing until they have a child who is *not* of the desired sex.
3. Plan 2 may be modified by the requirement that there be at least one child of the desired sex.

Plans may vary from couple to couple, complicating matters tremendously. We assume that the entire population follows the same plan and that boys are desired.

If p is the probability of success (or failure) in repeated Bernoulli trials, the expected waiting time until the first success (or failure) is

$$\sum_{n \geq 0} n(1 - p)^{n-1}p = p\frac{d}{dp}\sum (1 - p)^n = \frac{1}{p}.$$

Hence $E(N_i) = 1/p_i$ for plan 1 and $E(N_i) = 1/q_i$ for plan 2. For plan 3 we have either of the patterns boy(s)-girl or girl(s)-boy for order of birth of children. The first involves the birth of a boy and so has probability p_i and an expected number of births $1 + 1/q_i$. The second case is similar. Thus

$$E(N_i) = p_i\left(1 + \frac{1}{q_i}\right) + q_i\left(1 + \frac{1}{p_i}\right) = \frac{1}{p_iq_i} - 1$$

From this it is easy to compute approximations to μ by using (2). For plan 1 we obtain the harmonic mean of the p_i. Since the arithmetic mean exceeds the harmonic mean, this μ is *less* than random. This is due to the fact that high p_i is correlated with low $E(N_i)$. Similarly, plan 2 leads to a higher μ than random. What happens in plan 3 depends on the distribution of the p_i. Up to this point we have not needed any such information about the p_i. This is good, because they cannot be computed. See also page 217.

Making Simple Choices

What mental processes occur (possibly subconsciously) when you make a simple decision, like choosing the longer of two lines? No one really knows, and the models in this area are plagued by oversimplification; for example, a process can be assumed to be identical from trial to trial, or a relationship can be assumed to be linear, even though these assumptions are known to be only rough approximations. The following model, while no exception, illustrates some interesting ideas. It is adapted from R. J. Audley (1960). Another problem is the existence of several equally good (or bad) models for the same situation. See R. R. Bush and F. Mosteller (1959).

We wish to model an experimental situation in which a subject is required to choose between two simple alternatives, for example, which of two nearly equal lines is longer. The alternatives are called A and B, and the correct choice is A. We assume that the subject makes a sequence of choices implicitly (either consciously or subconsciously) and that these determine the final choice. Specifically, we assume

1. There are parameters α and β such that during a small time interval of length Δt implicit choice A occurs with probability $\alpha \Delta t$ and implicit choice B with probability $\beta \Delta t$. These events are independent.

2. A final choice is made after a run of K identical implicit choices, and it equals the implicit choice that was just chosen K successive times.

We consider only the two simplest cases of the model: $K = 1, 2$. It would be more appropriate to treat K as a parameter, but this would lead to more involved mathematics.

Assumption 1 implies that the next implicit choice is A with probability $p = \alpha/(\alpha + \beta)$. Let $q = 1 - p$. It follows that the probability of a string of implicit choices consisting of a A's and b B's in some given order is

$$(3) \qquad \text{Pr} \{a \text{ A's and } b \text{ B's}\} = p^a q^b = \frac{\alpha^a \beta^b}{(\alpha + \beta)^{a+b}}.$$

In an interval of length Δt,

$$\begin{aligned}
\text{Pr} \{\text{choice}\} &= \text{Pr} \{\text{A or B}\} \\
&= \text{Pr} \{\text{A}\} + \text{Pr} \{\text{B}\} - \text{Pr} \{\text{A and B}\} \\
&= (\alpha + \beta) \Delta t - \alpha\beta(\Delta t)^2, \qquad \text{by assumption 1.}
\end{aligned}$$

This describes what is called a *Poisson process* with parameter $\lambda = \alpha + \beta$. The properties of such a process are well known. In particular, the mean time between implicit responses is $1/\lambda$ and the probability of exactly n implicit responses during a time interval of length t is

$$(4) \qquad P_n(t) = \frac{(\lambda t)^n e^{-\lambda t}}{n!}.$$

The Poisson process is discussed in the Appendix. It also appears briefly in the radioactive decay example in Chapter 10.

We can use K, p, and λ as the basic parameters instead of K, α, and β, because $\alpha = p\lambda$ and $\beta = (1 - p)\lambda$. All the probability distributions can be parameterized by p and K if they are looked at as functions of $\tau = \lambda t$ rather than of t. Hence p and K determine the shape of distributions, and λ determines the time scale.

When $K = 1$, the subject makes only one implicit choice, and this is his final choice. The probability of a response by time t is $1 - P_0(t)$, which is Poisson by (4). Audley notes that this does not agree with experimental results.

When $K = 2$, the subject alternates between A and B in his implicit choices until he finally makes two identical implicit choices.

We begin by studying such strings of choices. Let $P_A(n)$ be the probability that there were exactly n implicit responses and the final choice was A, and let $P_A = \sum P_A(n)$ be the probability that the final choice was A. The probability of an n-long string ending in B is $(pq)^m$ if $n = 2m$ and $q(pq)^m$, if $n = 2m + 1$. (This allows for the case $n = 0$.) Hence, for $k > 0$, $P_A(2k) = p^2(pq)^{k-1}$, $P_A(2k + 1) = p(pq)^k$, and

$$(5) \qquad P_A = p^2 \sum_m [(pq)^m + q(pq)^m] = p^2 \frac{1 + q}{1 - pq}.$$

Since P_A can be determined from experimental data, we have a way of estimating p.

To estimate λ, some information involving time is required. Since means can usually be estimated fairly well from data, a mean time is a reasonable choice. Let L_A be the time to final choice given that the choice is A, and let L be the time to final choice regardless of whether it is A or B. As usual, we use notation like $\bar{L}$ to denote the mean of L, and $E(L)$ to denote the expected value of L. We have

$$E(L_A) = \frac{\sum n P_A(n)}{\lambda P_A}, \qquad E(L_B) = \frac{\sum n P_B(n)}{\lambda P_B}$$

$$E(L) = \frac{\sum n[P_A(n) + P_B(n)]}{\lambda}.$$

By using $\sum nx^n = x/(1 - x)^2$, it is an easy matter to evaluate these sums:

$$\sum n P_A(n) = \frac{p^2(2 + 3q - pq^2)}{(1 - pq)^2}$$

and so

$$(6a) \qquad E(L_A) = \frac{2 + 3q - pq^2}{(1 + q)(1 - pq)\lambda}$$

$$(6b) \qquad E(L) = \frac{2 + pq}{(1 - pq)\lambda}.$$

An interesting consequence of (6a) is that $r = E(L_A)/E(L_B)$ decreases from about $\frac{5}{4}$ to about $\frac{4}{5}$ as p increases from 0 to 1. To see this it suffices to study

$$f(p) = \frac{2 + 3q - pq^2}{(1 + q)(1 - pq)} = \frac{2}{1 - pq} + \frac{q}{1 + q},$$

since $r = f(p)/f(q)$. You should work out the details. Another way to describe the behavior of r is

The mean time to final choice is longer for the less likely choice, but it never exceeds the other mean time by more than about 25%.

We are now ready to compare the model with experimental results. Audley notes that very few suitable data are available and bases his major test of the model on the work of V. A. C. Henmon (1911). We rely exclusively on his work; see Audley's paper for further discussion. In his experiments Henmon displayed two vertical lines, one of which was slightly longer than the other. Half of the time the subject was required to choose the longer line, and half of the time, the shorter line. The subject was also asked to express a degree of confidence in the choice. The lines were displayed until a judgment was made. In a single series the subject was required to make 50 judgments. From three subjects 1000 judgments each were taken, and 500 each were taken from another seven subjects.

Unfortunately, only the data from the first three subjects is presented in a fashion that makes it possible to plot the number of decisions against time to decision (Henmon's Table II), the curve that would provide the most detailed test of the model. However, P_A, $\bar{L}$, $\bar{L}_A$, and $\bar{L}_B$ can be determined for all subjects by using his Tables I and IV. These are presented in Table 1.

Table 1   Choice Model Parameters for 10 Subjects

Subject	P_A	$\bar{L}$	$\bar{L}_A$	$\bar{L}_B$	$\bar{L}_B/\bar{L}_A$	$E(L_A)$	$E(L_B)$
Bl	0.820	1021	992	1154	1.16	1009	1079
Br	0.774	609	610	603	0.98	601	635
H	0.832	775	770	797	1.03	765	822
A	0.686	303	305	300	0.98	300	311
B	0.778	535	536	530	0.98	528	558
C	0.689	642	652	621	0.95	635	658
D	0.798	1044	1043	1046	1.00	1030	1096
E	0.778	1095	1046	1268	1.21	1081	1144
F	0.696	583	606	531	0.87	577	598
G	0.742	909	899	938	1.04	898	941

Note: Times are given in milliseconds.

We have taken A to be right and B to be wrong. Note that the value of $\bar{L}_B/\bar{L}_A$ for some of the subjects is less than 1, a contradiction to the theory. Some of the ratios are so close to 1 that the deviation is not significant, but the ratio for subject F is extremely low. Perhaps it can be explained by assuming that the value of K varied from series to series. You are asked to discuss this idea in the problems. After using P_A and $\bar{L}$ to estimate p and λ using (5) and (6b), the values of $E(L_A)$ and $E(L_B)$ were computed by (6a) and its analog for $E(L_B)$. Audley has fitted curves to the more detailed data (Henmon's Table II) for subjects Bl and Br. To do this he introduced a third parameter: a short time lag during which the subject prepares to make implicit decisions. It is then necessary to ignore the decisions made before the lag, because they occur before the subject is "ready." Without a time lag the fit is poor, but with a lag of 0.40 seconds for Bl and 0.34 seconds for Br the fit is good. I have not been able to obtain as good a fit for H as can be obtained for Br and Bl. Since there are so few data for each subject (four numbers), I think that the poor fit of the model is a sign of serious deficiencies; however, I'm not able to suggest a better model.

A related model has been proposed by Estes and Bower and extended by W. Kintsch (1963) to include a Poisson process for implicit response times. Assume there are five states: S, iA, iB, fA, and fB—starting, implicit A and B, and final A and B. The subject makes a decision to move from one state to another. The possibilities are

$$
\begin{array}{l}
 \quad \longmapsto \text{iA} \rightarrow \text{fA} \\
\text{S} \quad \updownarrow \\
 \quad \longmapsto \text{iB} \rightarrow \text{fB}
\end{array}
$$

Show that, if the probabilities of S → iA, iA → fA, and iB → iA are all equal, this reduces to Audley's model. Kintsch discusses primarily the case in which the probabilities of iA → fA and iB → fB are equal.

One problem that neither of these models deals with is the possibility of unconscious bias of the subject toward the right line or the left line. Another is the possibility that it is harder to choose the smaller than the larger, or vice versa. Either of these could lead to a mixing of models with different parameter values. Furthermore, data from different sessions with the same subject may have different parameter values. Any mixing like this could give rise to problems in fitting the data. Henmon's tabulations make it impossible to check all this out; however, he does note that there is a slight difference in reactions to the shorter line versus reactions to the longer.

PROBLEMS

1. Discuss the sex preference model when each couple can have no more than C children.

2. In this problem you'll consider ways of adapting Audley's model to fit Henmon's data more accurately. If you become very involved in this,

it would be a good idea to read Henmon's paper. Henmon obtained the following data for subjects Bl, Br, and H. He asked them to express a degree of confidence in their choice ranging from a (perfectly confident) to d (doubtful).

Confidence in choice	Subject Bl			Subject Br			Subject H		
	P_A	$\bar{L}_A$	$\bar{L}_B$	P_A	$\bar{L}_A$	$\bar{L}_B$	P_A	$\bar{L}_A$	$\bar{L}_B$
a	0.966	753	557	0.951	560	574	1	638	—
b	0.841	1045	987	0.944	596	669	0.972	722	699
c	0.653	1311	1205	0.836	635	606	0.853	789	777
d	0.480	1612	1499	0.615	624	596	0.563	850	814

(a) The simplest modifications of Audley's model may be either to choose a different fixed value for K or to allow p, λ, or K to vary while the other two are fixed. What do you think of this idea (before we actually examine it against the data)?

(b) Argue that, if p, λ, and K are all fixed, the accuracy of a decision depends only weakly on the speed with which it is made. How does this fit with the data? *Hint*: A decision corresponds to a mixture of A's and B's followed by K identical symbols (either A or B).

(c) Argue that P_A/P_B is approximately $(p/q)^{K-1}$ and that L is an increasing function of K and a decreasing function of p.

(d) Show that, if p and λ are fixed and K is variable, longer decision times are associated with greater accuracy. What if only λ varies? Only p? Which of these predictions seem reasonable in view of the data? Why?

(e) Can you propose a specific model which can be tested against Henmon's data? If you could have helped Henmon design his experiments, what would you have suggested he do differently in the actual running of the experiment and in the compilation of the data?

3. Develop the model of Kintsch, Estes, and Bower mentioned on page 98 with the equality assumption made by Kintsch. Compare the model with the data given above and in the text. Compare it with Audley's model with $K = 2$. Which seems to be better? Why? Can you suggest additional experiments that would be useful in testing the models?

4. Many colleges and universities are faced with a problem regarding tenured positions. To attract a good, young faculty, the prospects for tenure must be high, but to allow for adaptation, the percentage of

tenured positions should not be too high. What is the best strategy? The following material is adapted from an article by J. G. Kemeny (1973).

For our purposes let us distinguish three positions:

1, assistant professor (first appointment).
2, assistant professor (second appointment).
t, tenure.

Positions 1 and 2 each normally last for 3 years, and position t lasts for an average of about 30 years. Since these times are multiples of 3 years, we will take 3 years as the time unit. Let p_1 denote the probability of going from 1 to 2, p_2 the probability of going from 2 to t (given that the step from 1 to 2 has been made), and q_t the probability of leaving a tenured position (death, retirement, move to another institution) during a 3 year interval.

(a) Show that the probability of achieving tenure is $\tau = p_1 p_2$.

(b) Show that the fraction of faculty that has tenure in an equilibrium (i.e., steady state) situation is

$$\rho = \frac{p_1 p_2}{q_t(1 + p_1) + p_1 p_2}.$$

Hint: Let x, y, and z be the number of faculty in positions 1, 2, and t, respectively. Show that $E(y) = p_1 E(x)$ and $E(z) = (1 - q_t)E(z) + p_2 E(y)$.

(c) Conclude that, when τ is fixed, ρ is a minimum when $p_1 = 1$. Interpret this as a policy proposal.

(d) Kemeny estimates that q_t is roughly 0.15. Tabulate ρ versus τ for $p_1 = 1$. How sensitive is the tabulation to variations in p_1? Comment on the proposal in (c) in light of this.

(e) Incorporate appointments to the tenure level from outside and resignations from the assistant professor levels in the model. Hint: Look at flows of people as suggested in (b).

(f) Discuss the model. Is it realistic? Have important psychological factors been neglected? What psychological effect is the proposal in (c) likely to have on assistant professors? What would you recommend? Why?

5. In Section 3.2 the nuclear missile arms race was discussed qualitatively. This problem and the next one deal with a simple quantitative model discussed by T. L. Saaty (1968, pp. 22–25) and R. H. Kupperman and H. A. Smith (1972). See also K. Tsipis (1975). Suppose a country has M missiles which are being attacked by w warheads, each of which has a probability p of destroying the missile it is attacking. Suppose further that the behavior of the warheads is independent.

(a) Show that, if the ith missile is attacked by w_i warheads, $\sum w_i = w$ and the expected number of surviving missiles is

$$S = \sum (1 - p)^{w_i}.$$

(b) Show that the above expression is a minimum when the values of w_i are as nearly equal as possible. Interpret this in terms of strategy. Conclude that

$$S = M\left\{\frac{w}{M}\right\}(1 - p)^{1 + [w/M]} + M\left(1 - \left\{\frac{w}{M}\right\}\right)(1 - p)^{[w/M]}$$

$$= M(1 - p)^{[w/M]}\left(1 - p\left\{\frac{w}{M}\right\}\right)$$

$$\approx M(1 - p)^{w/M},$$

where $[x]$ is the largest integer not exceeding x and $\{x\} = x - [x]$ is the fractional part of x.

(c) Why is the variance of the expected value S important? Can you say anything useful about the value of the variance? With additional assumptions?

6. In the following discussion, use the results of the previous problem. To make the discussion uniform, assume that a retaliatory force of $S = 100$ surviving misssiles is desired and that $p = 0.5$.

(a) Suppose there are two equal countries (so $w = M$). Determine the minimum M required for stability.

(b) Suppose ABMs are installed to protect the defender's missiles. Why will this lead to a decrease in p? Plot M as a function of $p \le 0.5$. Discuss policy implications. Don't forget to take into account the limitations of the model. What if the attacker has ABMs that can protect its cities? (Consider S.)

(c) Suppose both countries introduce MIRVs with t warheads per missile. Discuss modifications in the formula for S and the desired value for S. It is fairly reasonable to assume that p is directly proportional to the cube root of the strength of the warhead and that this is proportional to the weight. It follows that $p(t) \approx p/t^{1/3}$. (Why?)

(d) Suppose there are three equal nuclear superpowers and each wishes to have a retaliatory force survive a coordinated attack by the other two powers. Discuss.

7. Have you ever noticed how children at a playground or people at a party form groups of various sizes? What sort of patterns are present?

This problem deals with the equilibrium size distribution of freely forming groups and was adapted from J. S. Coleman and J. James (1961). We assume that there is a collection of people who are free to join in groups as they choose. Examples are pedestrians, children playing, and shoppers. We wish to explain the size distribution of the groups. Five sets of data are given in the accompanying table. The first column

	I	II	III	IV	V
1	1486	316	306	.305	276
2	694	141	132	144	229
3	195	44	47	50	61
4	37	5	10	5	12
5	10	4	2	2	3
6	1	0	0	1	0

indicates the size of the group, and the remaining five columns refer to the five different groups observed by James. Data set I refers to pedestrians, data set II to shoppers, data sets III and IV to children at playgrounds, and data set V to people on a beach. The entries in the ith row are the number of groups of size i in each of the five samples.

(a) Let N be the total number of people present, G the total number of groups, and G_i the number of groups with exactly i members. Show that $G = \sum G_i$ and $N = \sum iG_i$.

(b) Suppose that in a very small time interval of length Δt single people (i.e., groups of size 1) join groups with probability $\alpha \Delta t$ per person, that the group joined is chosen at random, and that people leave groups and become single with probability $\beta \Delta t$ per person. Assume that people act independently of each other (in the probability theory sense of "independent"). Show that the expected net flow rate of groups from the collection of groups of size $i + 1$ to the collection of groups of size i is $\beta(i + 1)G_{i+1} - \alpha G_1(G_i/G)$ because groups of size $i + 1$ break up and groups of size i grow. Show that this must be zero at equilibrium, that is, although flow occurs, the *net* flow is zero.

(c) Let $p_i = G_i/G$. Interpret p_i and show that $\sum p_i = 1$. Show that at equilibrium $p_i = (p_1\alpha/\beta)^{i-1}p_1/i!$. Using this and $p_i = 1$, conclude that $p_i = \lambda^i/i!(e^\lambda - 1)$, where $\lambda = p_1\alpha/\beta$. (This is called a truncated Poisson.) Note that only the ratio α/β is important, rather than the actual values of α and β. Would this be true if we were concerned with a nonequilibrium situation? Why?

(d) We need a formula for λ in terms of the data. Show that $N/G = \lambda/(1 - e^{-\lambda})$ and use this to fit the model to the five examples given above. How good is the fit? (If you are familiar with the chi-square test, you may wish to use it.)

(e) Another way to fit the model is to estimate λ using $\lambda = (i + 1)p_{i+1}/p_i$; for example, $\lambda = 2p_2/p_1$. Is this a better idea than that in (d)? A worse idea? Why?

(f) Suggest further tests of the model besides the simple fitting of the data that you have done. Criticize the model. Can you justify proposing a model more complicated than the one developed here on the basis of the data? Why?

(g) Develop an alternate model by replacing "the group joined is chosen at random" in (b) with "the person associated with is chosen at random." Introduce $q_i = iG_i/N$ and $\lambda = q_1\alpha/\beta$. Show that $q_i = q_1\lambda^{i-1}$, $q_1 = 1 - \lambda$, and $G/N = (\lambda - 1)/\lambda \log (1 - \lambda)$. Which model provides a better fit to the data?

You may wish to look at J. E. Cohen (1971).

5.2. MONTE CARLO SIMULATION

When a probabilistic model cannot be analyzed analytically, *Monte Carlo simulation* is often used. The basic idea is to construct a deterministic model based on the probabilistic one by choosing particular values for the random variables according to the assumed distributions for them. Many such models are constructed, and statistical information is collected about the various dependent variables. This information is used to estimate parameters of the distributions of the dependent variables. If you don't have access to a computer, that's not reason to skip this section.

For example, suppose a "fair" coin is tossed 100 times. How many heads can we expect? The following is an algorithm for a Monte Carlo simulation of this problem.

1. Input N, the number of trials. Carry out steps 2 thru 4 N times.
2. Set HEADS to 0. Carry out step 3 100 times.
3. Choose X such that Pr $\{X = 0\}$ = Pr $\{X = 1\}$ = $\frac{1}{2}$. Set HEADS to HEADS + X.
4. Record the value of HEADS.
5. Analyze the data collected.

For this illustration, the analysis in step 5 will consist of determining the mean and variance of the number HEADS.

I ran the algorithm on a computer three times each for N = 10, 100, and 1000. The values of the mean and variance were:

N	Mean	Variance	Mean	Variance	Mean	Variance
10	51.2	14.8	49.1	14.5	49.6	38.2
100	49.7	26.2	49.6	21.5	49.0	21.4
1000	49.7	25.5	50.0	23.4	49.5	26.0

Note the greater variability in the estimates for the mean and variance when N is small. The theoretical values of the mean and variance are exactly 50 and 25.

How accurate are the estimates of the parameters of a distribution? Answering this question and obtaining more accurate estimates without an excessive number of trials are major problems in Monte Carlo simulation, but we only touch on them here. Given ε and δ greater than zero, we can obtain an estimate $\hat{S}$ of the parameter S such that

$$\Pr\{|\hat{S} - S| > \delta\} < \varepsilon,$$

provided the number of trials N is sufficiently large. Determination of N before simulation is usually very difficult; however, post hoc estimates can be made as follows. Assume that, when several estimates of S are obtained by simulation, they are drawn from a normal distribution with mean S. (This is probably not true, but often it is not too unrealistic.) If m estimates $\hat{S}_i$ have been obtained, $\hat{S} = \sum \hat{S}_i/m$ is a estimate of S and the variance of the estimate is given by

$$\sigma^2 = \frac{\sum (\hat{S} - \hat{S}_i)^2}{m(m-1)}.$$

If we apply this to the coin tossing problem we obtain the following estimates, the first $\hat{S} - \sigma$ pair referring to the mean and the latter referring to the variance. The value of m is 3.

	Mean		Variance	
N	$\hat{S}$	σ	$\hat{S}$	σ
10	50.0	0.6	22.5	7.9
100	49.4	0.3	23.0	1.6
1000	49.7	0.2	25.0	0.8
True	50		25	

The estimate for the mean happens to be the most accurate when $N = 10$. This is just chance; the best estimate we can give is 49.7. In addition to these ideas for measuring the accuracy of estimates, there is a theoretical result which can be used to obtain an idea of how many more trials we'll need: After N trials, the error in the estimate of a parameter is often roughly proportional to $1/\sqrt{N}$.

How can we generate the random choice required in step 3 of the coin tossing algorithm? Since a computer is (hopefully) a deterministic device, we cannot actually generate random numbers. However, almost every computer center has a subroutine which can produce a number between 0 and 1 each time it is called, and it does so in such a way that the entire sequence appears to have been sampled from the interval $[0, 1)$ using a uniform distribution. If a computer is not available, a table of random digits can be used: Simply start somewhere in the table, write a decimal point, and copy after it the next few digits in the table. This gives a random number drawn from the uniform distribution on $[0, 1)$. A brief table of random digits appears at the end of this chapter. Using uniformly distributed random variables, one can generate random variables according to any distribution. For example, if X is distributed uniformly on $(0, 1)$, the largest integer in kX is distributed uniformly on the set $\{0, 1, 2, \ldots, k - 1\}$. In general, if F is a distribution function, $F^{-1}(X)$ is a random variable with distribution function F. Since a table of F^{-1} can be constructed ahead of time, it is a relatively easy matter to choose random variables with the distribution function F. These ideas are discussed more fully in Section A.6. Here I'll content myself with two simple examples. The exponential distribution is given by $\Pr\{T > t\} = e^{-kt}$ for $t \geq 0$. Suppose $k = 2$. Then $F(t) = 1 - e^{-2t}$, and so $F^{-1}(x) = -\frac{1}{2}\log(1 - x)$. We generate five random values of T by using three-digit numbers from the table at the end of this chapter, starting with the first entry in the table:

X (table entry)	0.554	0.218	0.826	0.340	0.244
T (exponential)	0.404	0.123	0.874	0.201	0.140

Let's look at the uniform distribution on $\{0, 1, \ldots, k - 1\}$. In this case

$$F(t) = \begin{cases} 0 & \text{if } t < 0, \\ ([t] + 1)/k & \text{if } 0 \leq t \leq k - 1, \\ 1 & \text{if } t > k - 1, \end{cases}$$

where $[y]$ is the largest integer in y. Hence $F^{-1}(x) = [kx]$, as mentioned earlier. (There is a slight error in the definition of F^{-1} at points x for which

kx is an integer. Theoretically this is irrelevant, since these values occur with zero probability. Practically, the formula is correct because the uniform distribution comes from $[0, 1)$ instead of $[0, 1]$.)

A Doctor's Waiting Room

You've probably experienced a long wait for a doctor. Why does this happen? This problem is simple enough that a fairly realistic model can be analyzed theoretically using techniques of queuing theory. I plan to take advantage of the simplicity of the problem to work through a Monte Carlo simulation by hand, using the table of random numbers at the end of this chapter. On a normal day, Dr. Smock has his receptionist schedule one patient every 10 minutes from 9:30 A.M. to 11:50 A.M. and from 1:10 P.M. to 4:00 P.M., except that two patients are scheduled for 9:30 A.M. and no patients are scheduled for 10:40 A.M. or 2:40 P.M. Starting at 9:30 A.M. he works until all the morning patients have been treated, takes a lunch break, and then works until all the afternoon patients have been treated. Subject to the limitation that his lunch break is always at least 45 minutes, he sees the first afternoon patient at 1:10 P.M. or as soon afterward as possible. One week Dr. Smock's nurse was asked to time the patients' visits. She divided them into "short," "medium," and "long," according to the doctor's directions, and collected the data shown below.

Visit	Time Range (minutes)	Average Length (minutes)	Percentage of Total Visits
Short	3–7	5	38
Medium	7–15	11	47
Long	16–30	20	15

She also noted that the doctor spent 1 minute between patients and took 10 minute coffee breaks at 10:40 A.M. and 2:40 P.M., or as soon after these times as there was a break between patients. The receptionist observed that about 10% of the appointment times were not filled because of late cancellations and patients who failed to appear. Unfortunately, she did not notice if there was any bias toward certain times of day. No information is available on late arrivals, but the receptionist thought that patients usually arrived on time. That's the data we have to work with. Suppose we could have designed the data collection ourselves. What would you have asked for?

Before setting up the model it is interesting to note that according to the table Dr. Smock spends an average of 10 minutes with each patient he sees. Allowing for the 1 minute between patients and the 10% unfilled appointments, this works out to a full day for the doctor on the average.

Now we need to model the waiting room somehow. Various possibilities exist for modeling the amount of time a patient spends with the doctor. One of the simplest is to limit all visits to 5, 11, or 20 minutes each. Suggest others. I am going to use the following simulation and repeat it several times to generate data for several typical days. Criticize it and suggest improvements.

1. For each of the patient arrival times during the day, choose a random digit. If the digit equals zero, the patient doesn't arrive.
2. For each patient that arrives, choose a two-digit random number. If the number is at most 37, Dr. Smock sees the patient for 5 minutes. If the number lies between 38 and 84 inclusive, he sees the patient for 11 minutes. Otherwise he sees the patient for 20 minutes.
3. Using the results of the two previous steps and information about Dr. Smock's behavior we can put the doctor's day together.

I used the following method to model a day. On a sheet of paper for the day I had one row for each patient slot and five columns labeled "time in," "empty," "type," "see Dr.," and "time out." The "time in" column was filled with the various times allowed for appointments, namely, 9:30, 9:30, 9:40, 9:50, ..., 4:00. I then filled in the next column by reading a random digit from the table starting at the beginning of line 01 and using step 1. As a result, the 11:20, 1:20, 2:50, and 3:00 slots were empty. I then read the table two digits at a time to carry out step 2 for slots that were not empty. I obtained the following sequence of visits (short, medium, long, and—for empty): smsmmsmlsms–sml, lunch, m-lmssmsm--msmsss. As a result, the first 9:30 patient saw the doctor from 9:30 to 9:35, and the second saw him from 9:36 to 9:47, giving the 9:40 patient a brief wait. Continuing in this fashion to fill out the last columns, I found that the 10:50 patient didn't see the doctor until 11:07 because the doctor was running late and didn't have his coffee break until 10:56. As a result there were two patients in the waiting room very briefly at 11:10. The 11:20 cancellation allowed the doctor to catch up and even have a 3 minute break at 11:36. The afternoon was slightly slower, and the occurrence of two cancellations right after coffee break time allowed the break to run for 25 minutes.

To obtain some idea of how typical this was, I decided to model a second day. I picked up in the table of random numbers at the point I had left off at the end of the first day: the twenty-fourth entry on line 03. Although there was only one morning cancellation, things were a bit slow because of a large number of short visits. The afternoon was busier, with two patients in the waiting room twice, once for a quarter of an hour when the 3:40 patient arrived.

You may wish to model additional days and compare them with these two. If you think the waiting room tends to be rather empty and that the doctor would not like to have stretches during which he must wait for patients to arrive, you might like to adjust things by changing the scheduling. Scheduling two patients at times like 9:30 (already done), 9:40, 1:10, and 1:20 tends to build up a queue in the waiting room so that cancellations will not leave Dr. Smock at loose ends. You may wish to let the doctor work longer hours (an average of 30 minutes) to handle three extra patients, or you may wish to drop some appointments to make up for the additional ones, for example, 11:50, 3:50, and 4:00.

Sediment Volume

What happens when suspended particles settle? Do they attract each other? Slide after contact? It turns out that these things affect the density of the sediment. Thus we can obtain information about settling in an indirect fashion by studying the sediment's density. But how can such measurements be interpreted? We need a method for computing the density of the sediment under various assumptions. That's the purpose of this model.

We are interested in the fraction of volume occupied in a typical portion of the sediment, and we avoid the surface of the sediment where the fraction of volume occupied is not a well-defined concept. This model is adapted from M. J. Vold (1959, 1959a). For simplicity we assume that the particles in suspension are all spheres of the same size. Clearly the volume depends on whether the particles attract each other, cohere on contact, slide on contact, or repel each other. The last case can be eliminated, since we are assuming that the suspension settles. Which of the other cases occur? If attraction or sliding takes place, to what extent does it occur?

We cannot simulate the behavior of the entire suspension at once, but we can simulate the particles sequentially. Thus we can imagine a sediment into which we let particles settle one at a time. This may be a reasonable assumption if the suspension is fairly dilute. Discuss. Another problem we encounter is that in the real situation there are many more particles than we can possibly hope to include in the model. Although the model will have many fewer particles than a real life situation, it must have enough to avoid large random fluctuations and to avoid "edge effects" due to the bottom and sides of the container. After we propose the model, discuss whether you think there are enough particles. Can the question be answered by computation instead of on heuristic or philosophical grounds?

We treat the case of attraction and cohesion and leave sliding as a problem. Since the nature of the attractive force isn't specified, let's assume that it is zero when the distance between the centers of the particles exceeds

λr, r being the radius of a particle, and that it is infinite when the distance is less than λr and no other particle is closer to the settling particle. This is an unrealistic assumption, but it makes the modeling much easier and λ should give some measure of the attractive force. Discuss the effects of this crude assumption. When $\lambda = 2$, the model reduces to the case of cohesion. Why?

The Monte Carlo simulation proceeds as follows.

1. Choose a size and shape of cylindrical container, the radius r of the particles, and the number of particles. Repeat step 2 once for each particle.
2. Select a random point on the upper surface of the container and simulate a particle settling from this point until it comes to rest against another particle or on the bottom of the container. Record its location.
3. Gather the desired statistics.

The container is chosen to be cylindrical for simplicity. Since we are not interested in overflow, the container is chosen to be arbitrarily deep. We can easily set $r = 1$ and simply adjust the size of the container. Steps 2 and 3 require further explanation. The easiest way to keep track of a particle is probably by the location of its center, say with three coordinates (x, y, z), where x and y are in the horizontal plane and z increases upward. The point (x_0, y_0) at which a particle is dropped should be chosen randomly (this is the Monte Carlo part) by using uniform distributions on x and y. When a new particle is dropped, it ends up at a position (x', y', z'), determined as follows. For each previous particle with $(x - x_0)^2 + (y - y_0)^2 \le (\lambda r)^2$, find z_0 such that

$$(x - x_0)^2 + (y - y_0)^2 + (z - z_0)^2 = (\lambda r)^2$$

and choose the particle at (x, y, z) that gives a maximum z_0. Then (x', y', z') is the point on the line segment joining (x, y, z) and (x_0, y_0, z_0) that is a distance $2r$ from (x, y, z). If no particle is ever close enough, the new particle will settle to the bottom. You should convince yourself that this is correct.

The statistics we gather in step 3 will be the fraction of volume occupied by the particles. Since the upper surface of the sediment is not level, we take a cross section of the sediment well below the surface. This requires some numerical experimentation.

I chose particles of radius 1 in a container of such a shape that (x, y) for the centers of the particles would be in a square of side 14. Thus the cross-sectional area of the container is $A = 16^2 - 4 + \pi$, and the fraction of the volume occupied by n spheres in a section of container of height h is $4\pi n/3Ah$.

With 300 particles and $\lambda = 2$, I found that the volume fraction for $h = 10$, 15, 20, 25, 30, 40, and 50 was 0.163, 0.151, 0.147, 0.152, 0.150, 0.123, and 0.099, respectively. Hence it seemed reasonable to assume that $h = 20$ was well below surface effects but still large enough so that the volume fraction would not be much influenced by the flat bottom. I then made three runs each for various values of λ. Quite a while after these computations were done, N. P. Herzberg suggested that the possibility of difficulties with the bottom is indicated by the volume fraction for $h = 10$ and that these could be avoided by taking a slice between, say $h = 10$ and $h = 25$. Since the old program was gone, I decided to leave things as they were.

λ	Volume Fraction	σ
2.00	0.154530	0.009813
2.25	0.131340	0.003850
2.50	0.107535	0.013117
2.75	0.095632	0.003937
3.00	0.075931	0.008904

A downward trend in the volume fraction is quite visible. [Vold's simulations (1959) led to volume fractions slightly smaller than mine, but this may be due to the flat bottom.] She also determined the number of spheres contacting a given sphere and found that the average was very nearly 2. What does this mean?

Experimental results give a volume fraction of about 0.125 for glass spheres in nonpolar liquids and about 0.64 in polar liquids. How could this data be interpreted in terms of the models discussed here? (See also Problem 1.)

Stream Networks

Is there any regularity in stream networks? Some geomorphologists believe that many of the features of stream networks are random. In particular, are the branching patterns random? It would be nice to know, since if we found that they were non-random we could look for an explanation (or at least the geomorphologists could). What do we mean by "random" in this context? We use one idea of random adapted from A. E. Scheidegger (1970, Sec. 5.33).

First we need some definitions. A drainage basin consists of a *stream* (*or river*) *network* and the area it drains. A stream network is a stream together with all the streams that flow into it above the point at which we are considering the stream. A *link* is the portion of a stream between two junctions or between a junction and a source. A stream network is almost always made up of a set of links joined so that at each junction only two links flow together to form a third. (The rare occasions when more than two streams

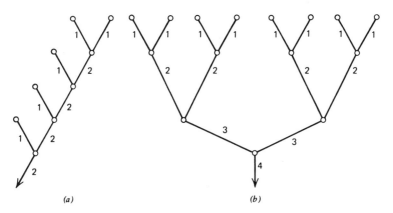

Figure 1   Two extreme examples of stream networks.

meet simultaneously can be resolved, but we won't go into that complication here.) See Figure 1. The *Strahler order* of a stream link is defined as follows. Links that start at a source are of order 1. If two links of orders A and B flow into a third link of order C, then C equals $A + 1$ if $A = B$, and C equals the maximum of A and B otherwise. See Figure 1. A *segment* is a stretch of river over which the order doesn't change. Let n_i be the number of segments of order i. Thus $n_2 = 1$ in Figure 1a and $n_2 = 4$ in Figure 1b. Horton's law of stream numbers is an empirical relationship which states that n_i/n_{i+1} is nearly independent of i. For streams in the United States, this approximate constant (whatever that means) is about 3.5 according to Scheidegger. However, the data of L. B. Leopold et al. (1964, p. 142) for the entire United States, presented in Table 2, does not agree with this. If stream networks tend to be fairly linear as in Figure 1a or rather bushy as in Figure 1b, this law is not valid. (Compute n_i and n_i/n_{i+1} in these cases.) It has been suggested that the result can be explained by assuming that stream networks are random. We model this idea following Liao and A. E. Scheidegger (see A. E. Scheidegger, 1970).

The only geometric property of a stream network we have introduced is the pattern of connection among the links; lengths and curvatures have been omitted. Given the number of sources, there is only a finite number of different drainage networks. Those with four sources are shown in Figure 2. These patterns of connection are known mathematically as *plane planted binary trees* ("trees" because of shape, "plane" because they are drawn on a flat surface, "planted" because the link at which we have cut the network is distinct from all others and can be used to plant the tree, and "binary"

Table 2   Number of Stream Links of Various Orders in the
United States.

Order	Number	Average length (miles)	n_i/n_{i+1}	Example
10	1	1800	—	Mississippi
9	8	777	8.0	Columbia
8	41	338	5.1	Gila
7	200	147	4.9	Allegheny
6	950	64	4.8	
5	4200	28	4.4	
4	18000	12	4.3	
3	80000	5.3	4.4	
2	350000	2.3	4.4	
1	1570000	1	4.5	

Source: L. B. Leopold, et al. (1964).

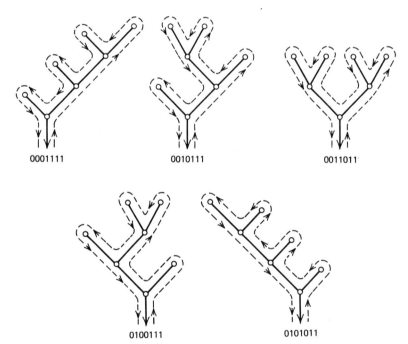

Figure 2   The 5 seven-node plane planted binary trees and their seven digit Lucasiewicz sequences.

because of the bifurcation at each node as we move upstream). Since this is the only type of tree we care about here, we call them simply "trees." It is not hard to show that a tree with n sources has $2n - 1$ nodes and $2n - 1$ links (including the link at which we have cut the network for study, that is, the link furtherest downstream).

To study n_i/n_{i+1} we want to average over all trees with n sources, or at least over a reasonable number of randomly generated n-source trees; that is, each of the trees with n sources is equally likely to be chosen. Since Horton's law is formulated for stream networks of fair size, we want n to be fairly large. When n is about 100, there are about 10^{56} trees—far to many to generate all of them. Thus we need a way to generate and store a random tree in a digital computer. Fortunately this mathematical problem has a fairly simple solution due to Lucasiewicz. We imagine traveling along the tree so that each link is traversed exactly once upstream and exactly once downstream. We start upstream on the link used to plant the tree, use the following rules, and stop when we return downstream on the cut link.

1. Go upstream if possible.
2. If a choice is possible, go upstream on the right hand branch.
3. When a node that is not a source is encountered while going upstream on a right hand branch, record a zero.
4. When a source is encountered, record a one.

This process is illustrated in Figure 2. It is possible to reconstruct the tree from the string of zeroes and ones:

1. Draw the planted link.
2. If the next digit is a zero, draw a bifurcating node and proceed upstream on the right hand branch.
3. If the next digit is a one, draw a source and proceed downstream until an untraversed upstream link is found. Go up it.

You should convince yourself that this algorithm does indeed work.

A string of zeroes and ones corresponds to a stream network with n sources if and only if it possesses two properties:

1. The number of ones in each initial segment never exceeds the number of zeroes.
2. The total number of ones equals n, and the total number of zeroes equals $n - 1$.

The second requirement just says that there are n sources and $n - 1$ internal nodes. The first requirement ensures that as we go downstream we never return to the link at which we have cut the network before the last step. Since properties 1 and 2 are necessary and sufficient and since all trees are obtained exactly once in this way, it suffices to generate sequences satisfying properties 1 and 2 randomly. A method for doing this is discussed in Problem 2.

Given an internal representation of a tree, we need a way to find n_i. This can be done as follows. We list the nodes in the order first reached by traveling around the tree as described above. Each node refers to the link immediately downstream from it. Construct two sequences, LORDER and ORDER: the first refers to the order associated with the left hand branch and the other refers to the actual order. We proceed in order through the sequence L of zeroes and ones which represent the tree. If $L_I = 0$, do nothing if $L_I = 1$:

1. Record 1 in ORDER_I and LORDER_I and set ORDERNOW to 1.
2. Find the nearest preceding LORDER_J which is blank (i.e., $J < I$, J is a maximum, and LORDER_J is blank) and do the following for $K = I - 1$, $I - 2, \ldots, J + 1$.

   a. Record in each blank ORDER_K the maximum of LORDER_K and ORDERNOW if $\text{LORDER}_K \neq$ ORDERNOW and record $1 + \text{LORDER}_K$ if $\text{LORDER}_K =$ ORDERNOW.
   b. Set ORDERNOW equal to the value of ORDER_K just recorded.

3. Set LORDER_J equal to ORDERNOW.

Work your way through some examples and try to see why this method works.

Note that in this Monte Carlo simulation the main problem is constructing algorithms for handling the pictorially simple concepts of tree and order in a digital computer. We have one problem left: How do we identify segments? This is fairly easy. When we are computing the order of a link, it will be a new segment if it is a source, or if the orders of both branches feeding in are equal; otherwise it will belong to a segment containing either the left or right branch. It is useful to keep a sequence SEGMENT that notes which links are the furtherest upstream link of some segment.

I generated random stream networks using the above ideas and found a result similar to that obtained by Liao and Scheidegger: For fixed i, the value of n_i/n_{i-1} increases slowly with n to about 4.0. When $n_{i-1} \geq 15$, the expected value of the ratio appears to exceed 3.8. Do you think this is evidence in favor of the random stream network model or against it? Why?

Can you suggest other tests? See A. E. Scheidegger (1970, Ch. 5) for further discussion.

S. B. Barker et al. (1973) made some studies of the branching structure of real trees. They counted all the branches on an apple tree and on a birch tree. For the apple tree they found that n_i/n_{i-1} was about 4.35, and for the birch tree it was about 4.00. Does this look random?

It would be a good idea to try a different approach to the idea of what a random network is, if we can think of one. One possibility is discussed in the problems. M. J. Woldenberg (1969) discusses yet another approach to understanding stream networks and criticizes the claim that n_i/n_{i-1} is independent of i. His method is an adaptation of the geoeconomic marketing model called *central place theory*. See S. Plattner (1975) for a discussion.

Trees and other graphs are useful tools for some types of modeling problems. You may enjoy reading F. S. Roberts (1976, Ch. 3).

PROBLEMS

1. Construct a Monte Carlo simulation model for sediment volume when the particles are allowed to slide downward in settling. Can you explain the volume fraction for polar solvents by this model?

2. We want to choose sequences of zeroes and ones satisfying properties 1 and 2 in the stream network example.

    (a) Show that, if a sequence satisfies property 2, exactly one "rotation" of it will satisfy property 1. A rotation of $d_1, d_2, \ldots, d_m$ is a sequence $d_{1+i}, d_{2+i}, \ldots, d_{m+i}$, where $d_j = d_k$ with $1 \le k \le m$ and $j - k$ a multiple of m.

    (b) Use (a) to construct an algorithm for rotating a sequence satisfying property 2 to obtain one that satisfies property 1.

    (c) We now want an algorithm for randomly choosing k positions from m in such a way that each of the possibilities is equally likely. Find one.

    (d) Combine the above to produce a complete algorithm for randomly generating strings of zeroes and ones that represent trees.

3. A manufacturing plant is trying to decide whether to increase the number of loading docks for trucks. Truck arrival at the docks is *not* uniform during the working day.

    (a) Describe how you would set up a Monte Carlo model to help management decide how many loading docks to have. Remember that it must be reasonable to collect the data. You should work the model out to the point where you could carry out the simulation if data were supplied.

(b) Discuss in class what factors could lead to nonuniform arrival rates. Choose a specific situation that leads to nonuniformity and hypothesize some reasonable arrival rates. (Note that for the number of docks to be about right, as it presumably is, the number of arrivals per day should average somewhat less than the loading docks could handle by working steadily. Why?) Choose a particular Monte Carlo simulation method from (a), hypothesize reasonable data, divide up the work, and do the simulation by hand. During the next class period pool your results so as to answer management's question.

4. How many comets are there in the solar system? What is the rate of loss of comets from the solar system? The following model deals with the number of "long period" comets in the solar system and follows J. M. Hammersley (1961). An interesting feature is that, although we usually think of the laws of planetary motion as a classic example of a deterministic system, Monte Carlo simulation is useful. This is because the number of comets is large. We had a similar situation in the sedimentation problem.

A long period comet is a comet that goes well beyond the orbit of Jupiter, and by "comet" we mean a long period comet. If we measure the energy E of an object orbiting the sun in such a way that it is zero when resting at an infinite distance, by one of Kepler's laws, the period T of the orbit equals $(-CE/m)^{-3/2}$, where m is the mass of the object and the constant C depends only on the gravitational constant and the mass of the sun. If $E \geq 0$, the object will escape from the solar system.

(a) What can cause E to change? The main influence is the gravitational field of Jupiter. Discuss others. If we set $z_i = -CE/m$, where E is the energy after the ith pass by Jupiter's orbit, Δz_i can be treated as a random number with a distribution depending on Jupiter and the sun but not on m. Approximate this by a normal distribution with mean zero. How could you check this approximation? [See R. H. Kerr (1961).]

(b) Show that, up to scaling, the lifetime of a "random" comet is given by

$$\sum_{i=0}^{T-1} z_i^{-3/2},$$

where $z_i > 0$ for $1 \leq i \leq T - 1$, $z_T \leq 0$, and Δz_i has a normal distribution with mean zero and variance one. What is the scale factor?

(c) Describe a Monte Carlo model for obtaining information about the distribution of lifetimes of comets, when time and z_0 are measured in whatever units were necessary for scaling.

(d) If most comets wander into the solar system from outside, as is believed by some astronomers, what is a reasonable value for z_0? Should we neglect $z_0^{-3/2}$ in (b)? Why?

(e) How could we estimate the total number of comets in the solar system, assuming losses and gains are equal and (d) holds? Hammersley obtained an estimate of about 2 million comets.

(f) Suppose all comets were formed within the solar system when it came into being. Discuss changes in (d) and (e).

5. We consider another way to approach randomness in stream networks. The idea is that the topography is random. Imagine a portion of a plane covered with squares. We think of the edge of each square as a possible stream link. Water might flow from or through any given vertex to an adjacent vertex. See Figure 3. This idea was suggested by a discussion in L. B. Leopold et al. (1964, p. 419).

(a) Given a vertex v, choose an adjacent vertex at random and allow the water to flow from v to w. Be careful. We can't do this if we've previously decided to let water flow from w to v. Bifurcating sources and "lost" rivers must be avoided. See A and B in Figure 3. How could you implement this on a computer? What about the possibility of water flowing in a closed loop such as C in Figure 3? Can you handle this by allowing lakes or by somehow stopping it by clever programing?

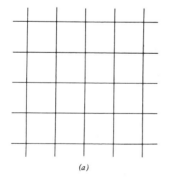
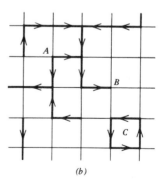

(a)   (b)

Figure 3 Choosing random stream networks on a grid. (a) Portion of grid. (b) Randomly generated links on this portion of grid. Problems have arisen at A, B, and C.

(b) Change the model in (a) so that each vertex is assigned an altitude and water runs down to the lowest adjacent vertex. What problems arise in implementation?

(c) Discuss biasing the two models just suggested to allow for a general overall slope to the land.

(d) Since four edges meet at the vertex of a square, we can expect to have some vertices where three streams join to form a fourth. This can be avoided by using a hexagonal (honeycomb) pattern instead of a square pattern.

(e) Criticize the model.

(f) Perhaps some students can actually implement a Monte Carlo model. If this is going to be done, discuss the practical details carefully beforehand. Among the things you will need to consider are:

Which model(s) will be implemented?
How should the model be stored?
How big should the model be?
How can the order of a link be determined?
How can the segments be identified?
Exactly what data, if any, are needed?

Don't forget the problems mentioned in (a) and (b).

A Table of 3000 Random Digits

01	55421	88263	40244	60613	18750	09668	67045
02	21661	65304	89606	67132	56488	75977	93311
03	77254	57610	76372	92693	08168	45645	96331
04	03803	63025	94237	33227	51828	07254	96652
05	29005	68581	18068	71414	93529	03790	17147
06	90086	72725	85496	36015	19475	79306	88066
07	48786	42078	66302	79185	47917	31532	59264
08	01312	06015	96224	42768	22830	78005	17433
09	90897	96649	85718	42458	18222	68868	36204
10	11433	10412	53251	08366	26673	89379	27952
11	74500	34547	78695	98961	50370	12118	80601
12	01710	94533	38266	42999	85821	12617	98876
13	98325	93297	87417	79283	13082	73321	08108
14	91318	54562	90536	39274	26757	04007	76649
15	65640	33035	47348	50884	71729	31237	96000
16	33578	71492	89085	24821	58763	03745	50706

17	38934	90627	93619	12976	74853	36562	52889
18	63994	37135	04933	28191	71590	16916	89009
19	56799	35247	78481	70048	75596	96136	09513
20	12726	49439	33920	67668	25313	05208	07753
21	34081	69899	92802	81144	52246	20404	66428
22	83547	15593	24422	56988	07032	16541	80267
23	62794	95699	55102	57232	04292	24619	00792
24	95447	93642	41265	11687	85266	95769	85657
25	26596	38328	75787	79328	64024	81217	14914
26	74519	73834	73701	61159	75618	10719	23249
27	76702	12394	98323	11486	65591	66169	61371
28	93398	25450	41967	89708	93328	08532	17663
29	03921	70788	45139	50713	83241	46227	81250
30	07876	78832	93503	46088	28554	49913	56826
31	17597	12602	71925	63115	51767	13525	65363
32	28348	46747	05225	11003	99959	69238	13750
33	57790	22390	75625	05258	14261	27013	10094
34	13233	96412	29753	95187	60401	53309	16058
35	35809	47147	66631	87135	39573	98117	12344
36	99902	47164	61113	79916	65611	28481	05621
37	12851	76785	25019	79805	01740	68627	82308
38	87584	17122	15362	56795	18723	54025	13867
39	21627	33387	94307	34270	22996	79509	97534
40	39124	97154	28543	15167	98577	22030	37310
41	83985	65741	00115	66382	02337	01885	26932
42	72642	09689	88779	68543	64174	27344	38379
43	86351	00215	97630	62359	24386	52426	87404
44	78675	13948	23670	20818	41693	69965	45507
45	13744	07743	55507	62664	04571	78498	05944
46	71582	87153	45222	95055	30583	88348	92666
47	80380	39093	97093	68003	00416	76429	04361
48	99964	70393	24149	23608	58032	39520	16090
49	05032	42931	69890	80165	13916	71993	25752
50	58991	74921	38536	68391	72232	85406	95680
51	07667	26870	48732	42076	86542	33490	49293
52	40078	77005	00604	53344	21916	31700	72849
53	30787	46512	89824	81494	04148	74399	03683
54	27095	59999	79940	23254	28226	46871	11524
55	47394	01133	87725	45405	91783	60142	24679
56	64478	56998	91421	61692	83308	23590	73162
57	20095	81826	77211	42919	56828	53315	23430
58	29785	41130	48891	69755	06426	33279	89180

59	06122	50707	70290	74073	82102	40049	49514
60	95826	83455	41687	28490	31137	55658	19873
61	33419	47261	13998	42627	70392	75443	75939
62	13127	42437	39921	97912	60053	75764	04210
63	17970	44384	95134	01034	51693	83968	41619
64	02440	09677	25867	50480	55276	39445	86379
65	01902	33280	69006	57137	75395	58215	16067
66	83708	61287	95269	63918	66823	85887	47487
67	43366	45811	45506	02740	12387	35925	69605
68	28400	81384	56051	49615	17959	91881	07447
69	10878	67992	50896	20390	28689	02029	27049
70	35304	33948	64811	09205	00181	59797	53427
71	74794	04070	13049	78158	40274	18380	31390
72	50612	11495	56502	37454	15523	17100	29111
73	90297	95935	31036	83853	91422	14307	66632
74	07048	79736	76495	68263	22727	72509	52840
75	70827	68071	70123	09804	84209	64910	73477
76	34161	49740	02489	00271	66229	66429	53530
77	13889	95558	55047	99000	21703	34104	03878
78	90726	42834	45339	56711	56299	35935	45020
79	54383	76347	29876	19497	84310	96346	51867
80	94345	29276	07885	14461	64927	41423	09201
81	72425	54109	47783	67259	68498	69107	15027
82	79981	59796	78249	05050	68335	25702	25771
83	83129	35323	59702	12961	22452	71264	86662
84	09583	04316	57908	37926	10256	73089	79661
85	52392	87142	65066	58787	76981	91372	72138
86	66641	47752	48858	56250	61530		

CHAPTER 6

POTPOURRI

The models presented here use a variety of elementary methods that didn't fit conveniently into the earlier chapters.

Desert Lizards and Radiant Energy

Lizards in arid regions make use of radiant energy (direct sunlight, reflected sunlight, and infrared radiation from the ground), conduction of heat through contact with the ground and with rocks, and convection to adjust their body temperature. Because of the high reflectivity of the sand (about one-third of the sunlight is reflected) and the heat of the sand, one could suppose that reflected sunlight and infrared radiation are nearly as important as direct sunlight. This model, which is adapted from K. S. Norris (1967), studies the question.

Since we wish to compare the relative amounts of energy hitting the lizard, its actual shape is not likely to be very important. Since symmetry usually simplifies computations, we assume that the lizard is a sphere of radius r whose center is a distance h above the sand. We assume that the sun is directly overhead and has an energy E per unit area per unit time. We consider the ratio of reflected sunlight to direct sunlight.

The energy per unit time due to direct sunlight is $\pi r^2 E$.

To study the reflected light we take advantage of the symmetry by setting up a polar coordinate system on the sand with its center directly below the lizard. A side view is shown in Figure 1. The fraction of light reflected from the sand at point P that reaches the lizard depends on the distance ρ, the angle φ, and the angular diameter of the lizard as seen from P. As a first approximation, let's suppose that the intensity of the reflected light is independent of φ. Then the fraction of light hitting the lizard will nearly

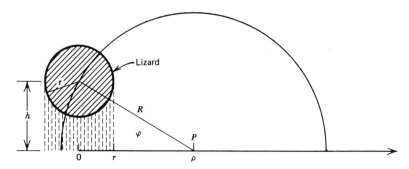

Figure 1 Side view of a spherical lizard at high noon.

equal the fraction of the hemisphere of radius R centered at P that lies within the lizard. This fraction is nearly

$$(1) \qquad \frac{\pi r^2}{2\pi R^2} = \frac{r^2}{2(\rho^2 + h^2)}.$$

The total amount of sand surface between ρ and $\rho + d\rho$ is $2\pi\rho\, d\rho$. Since the area directly under the lizard is shaded and since about one-third of the incident light is reflected, it follows from the above discussion that the amount of reflected light reaching the lizard from the sand up to a distance x away is nearly

$$(2) \qquad \int_r^x \frac{r^2}{2(\rho^2 + h^2)} \frac{E}{3} 2\pi\rho\, d\rho = \frac{\pi r^2 E}{6} \log\left(\frac{x^2 + h^2}{r^2 + h^2}\right).$$

Dividing this by the direct energy we obtain

$$(3) \qquad \frac{\text{Reflected}}{\text{Direct}} = \frac{1}{6} \log\left(\frac{x^2 + h^2}{r^2 + h^2}\right).$$

As x becomes large, (3) approaches infinity. What is wrong?

One objection is that we have treated the desert as a flat, barren plain, which is certainly not correct. Suppose that topography and brush begin to interfere seriously with the reflected light at a distance between $5r$ and $500r$. If h is at most $2r$, the value of (3) will be between 30 and 200% for x between $5r$ and $500r$. This answer appears to be quite reasonable, and there is no need to determine very accurately when brush and topography become important unless we want very accurate estimates of (3).

A completely different objection is that the intensity of reflected sunlight does indeed depend significantly on the angle of reflection. "Significantly" may be rather misleading here, since at a distance of about $500r$ we have $\varphi = 0.2°$ and so no reflection at angles less than $0.2°$ is sufficient dependence to limit reflected sunlight to a reasonable value. Let's consider the general situation.

We can allow for dependence on the angle by introducing a function $f(\varphi)$ multiplying the integrand in (2). This function should vanish at $\varphi = 0$ and achieve a maximum of 1 at $\varphi = 180°$. I have been unable to find an empirical estimate of f. One possible function is the sine. It has the right general form and leads to an integral which can be easily evaluated:

$$
(4) \qquad \int_r^x \frac{r^2 h}{2(\rho^2 + h^2)^{3/2}} \frac{E}{3} 2\pi\rho \, d\rho = \left. - \frac{\pi E r^2 h}{3(\rho^2 + h^2)^{1/2}} \right|_r^x
$$

$$
\rightarrow \frac{\pi E r^2 h}{3(r^2 + h^2)^{1/2}} \qquad \text{as } x \rightarrow \infty.
$$

Thus we have

$$
(5) \qquad \frac{\text{Reflected}}{\text{Direct}} \leq \frac{h}{3(r^2 + h^2)^{1/2}},
$$

which is bounded above by $\frac{1}{3}$. This result is of the same order of magnitude as the result obtained previously. We could consider other forms for f and other values for x. In the end we would probably find that for anything reasonable the ratio of reflected to direct sunlight was at least 20%—a significant amount of energy. Infrared radiation probably behaves in a similar fashion, hence reflected sunlight and infrared radiation are important factors in a lizard's heat balance.

Attempts have been made to use these crude results to study what happens as parameters vary, but this can be dangerous. To see this let's consider what happens when a lizard adjusts h by bending its legs. By differentiating with respect to h it is easy to see that the right hand side of (3) is a decreasing function of h and the right hand side of (5) is an increasing function of h. Thus our model is not good enough to tell us whether the lizard becomes warmer or cooler when it raises itself. Actually the lizard will probably become cooler because of an important effect that has not been mentioned: A thin layer of hot air is found on the surface of the sand. If you wish another example of the difficulties that arise from not knowing f, consider the following. Will a lizard in a bowl-shaped depression in the sand be warmer or cooler than an identical lizard on the flat sand?

Are Fair Election Procedures Possible?

In a mathematical model we normally use mathematics to study approximately the behavior of a real situation. In this example we consider a different type of question: What can we deduce about a situation that satisfies certain conditions? This is the axiomatic approach of pure mathematics: Make certain assumptions and see where they lead. The problem is to choose reasonable assumptions which lead to interesting conclusions. One of the earliest and most successful examples of the axiomatic method in science is Newtonian mechanics. This approach has also been used in sociology and economics. A particularly successful example is utility theory. See R. D. Luce and H. Raiffa (1958) for a discussion. J. F. Nash (1950) applied the theory to show that with some additional axioms one can conclude that there is a unique "fair" trade in two-person bargaining. J. G. Kemeny and J. L. Snell (1962, Ch. 2) showed how certain axioms lead to a unique measure of the distance between individual preferences. Here we study elections. Our goal is to prove that there is no fair way to run an election between several candidates. This is known as the *Arrow impossibility theorem*. This version differs slightly from that of K. J. Arrow (1962, Ch. 8). I've selected this particular example because it is easy to present, is somewhat surprising, and conveys the flavor of the axiomatic method. For a discussion of these topics see F. S. Roberts (1976, Chs. 7 and 8).

We need to say what we mean by a fair election procedure; but before we can do that, we must say what an election is. Letters like x, y, and z denote candidates, and letters like i and j denote voters. A *ranking* (also called an ordering) is a relation $\supset$, read "is preferred to," satisfying

1. For all x and y, exactly one of $x \supset y$, $y \supset x$, and $x = y$ (read "x and y are tied") is true.
2. For all x, $x = x$.
3. For all x, y, and z, if $x \supseteq y$ and $y \supseteq z$, then $x \supseteq z$ with $x = z$ if and only if $x = y$ and $y = z$.

We assume that each voter has ranked the candidates, and we use $(x \supseteq y)_i$ to denote the ranking given by voter i. An *election procedure* is a rule for deducing a ranking, denoted simply $x \supseteq y$, from all the individual rankings. Note that an election is not just a choice of the top candidate, but rather a ranking of all the candidates. If the procedure is fair, we will obtain a complete ranking from a procedure that gives the top candidate; for example, to find the second ranking candidate we apply the procedure to find the top candidate we apply the procedure to find the top candidate when the winner is removed. This can be formally justified on the basis of axiom 3 below.

Rather than specify exactly what constitutes a fair election procedure, I'll list some conditions (axioms) an election procedure must satisfy if it is fair. You may wish to add others, but you are not allowed to remove any of the following five. After listing them, I'll discuss them.

1. All conceivable rankings by the voters are actually possible.
2. If $(x \supseteq y)_i$ for all i, then $x \supseteq y$ with equality if and only if $(x = y)_i$ for all i.
3. If in two different elections each voter ranks x and y the same, then the election outcomes between x and y are the same; that is, if for all i $(x \supseteq y)_i$ if and only if $(x \geq y)_i$, then $x \supseteq y$ if and only if $x \geq y$. Here $>$ denotes the other election.
4. If there are two elections such that $(x \supseteq y)_i$ implies $(x \geq y)_i$ for all i, and if also $x \supseteq y$, then $x \geq y$.
5. There is no i such that invariably $x \supseteq y$ if and only if $(x \supseteq y)_i$.

The first condition says that the election procedure must be able to deal with all cases. The second axiom simply states that a unanimous desire of the voters is respected by the election procedure. Axiom 3 says that how two candidates rank relative to each other in the election depends only on how the voters rank them relative to each other and not on how they rank relative to other candidates. Thus inserting other candidates won't change the election ranking of x relative to y. Axiom 4 states that, if x does at least as well compared to y in a later ranking by the voters as he did in the present ranking, and if he beat y in the present election, he'll beat y in the later election. In other words, if your relative position improves in the eyes of all the voters, it will improve in the election results. The final assumption says that there is no dictator.

We can manipulate these axioms in a variety of ways to reach conclusions. In fact, it can be shown that axiom 3 follows from the rest. (You might like to try to prove this.) The manipulations we are interested in are those that lead to a proof of the following impossibility theorem.

THEOREM. No election procedure for more than two candidates satisfies axioms 1 through 5. Hence a fair election procedure is impossible if there are at least three candidates.

PROOF. We show that axioms 1 through 4 imply that there is a dictator. Note that, if we have an election procedure for N candidates, we can obtain one for $N - 1$ candidates by introducing a dummy Nth candidate which all the voters are assumed to rank lowest. It is easy to show that, if assumptions 1 through 4 hold for the original procedure, they hold for the derived procedure.

A set of voters V will be called decisive for x against y if when all voters in the set V agree on ranking x at least equal to y, then $x \supseteq y$ regardless of how the remaining voters rank x and y; furthermore, we require that in this case $x = y$ implies that $(x = y)_i$ for all i in V. At least one decisive set exists for all x and y—all the voters are decisive by axioms 1 and 2. Note that by axiom 4 we can check if a set is decisive just by looking at an election with $(x \supseteq y)_i$ for all i in V and $(x \subset y)_i$ for all i not in V.

We show that for some x and y there is a single voter who is decisive. Suppose that this is not true and let V be the smallest decisive set. Then V has at least two voters in it, and so we can split it into two nonempty, disjoint sets of voters V_1 and V_2. Let z be another candidate and consider an election in which

$$(x \supseteq y \supseteq z)_i \quad \text{for} \quad i \text{ in } V_1,$$
(6) $$\quad (z \supseteq x \supseteq y)_i \quad \text{for} \quad i \text{ in } V_2,$$
$$(y \supset z \supset x)_i \quad \text{for} \quad i \text{ not in } V.$$

If $x \supseteq z$, then V_1 is decisive for x and z, contradicting the minimality of V. Thus $z \supset x$. Since V is decisive for x and y, it follows from (6) that $x \supseteq y$. Thus $z \supset y$. Hence V_2 is decisive for z and y, contradicting the minimality of V. (One has to be careful to check out the cases where equality occurs. I won't bother because it clutters up the proof and I only want to give you the flavor of this type of argument.) Thus V contains a single voter, say i.

We have shown that for the two candidates x and y, if $(x \supseteq y)_i$, then $x \supseteq y$. Let z be a third candidate. Now suppose that $(x \supseteq y \supseteq z)_i$. Consider the election when $(y \supset z \supset x)_j$ for all $j \neq i$. By axiom 2, $y \supset z$, and by decisiveness, $x \supseteq y$. Hence $x \supset z$. By axiom 3 we can ignore y and note that, if $(x \supseteq z)_i$ and $(z \supset x)_j$ for all $j \neq i$, then $x \supseteq z$. Hence i is decisive for x and z. Let w be a candidate distinct from x and z. By a parallel argument we can show that i is decisive for w and z. This shows that i is decisive for every pair; that is, i is a dictator.

This completes the proof.   ■

How does this work out in practice? Suppose a contract administrator sends contract proposals (candidates) to experts (voters) for ranking and then determines a final ranking (election). Although he may not weigh the opinions of the experts equally, we hope that his ranking procedure will be fair. The theorem says that this is impossible, and the administrator may not actually be aware of this fact. What axiom is he violating? It is unlikely to be either 2 or 5. Since 3 follows from the other axioms, he must be violating 1 or 4. In other words, either the administrator cannot produce a ranking in all cases (such situations could be handled by obtaining additional voters) or the ranking of other proposals influences how he decides to rank proposals x and y relative to each other.

Impaired Carbon Dioxide Elimination

It is relatively easy to measure the concentrations (via partial pressures) of various gases in the air exhaled and inhaled by a person. Thus this could lead to a diagnostic test—if we know how to interpret the data. In 1922 Haldane asserted that carbon dioxide (CO_2) elimination by the lungs is generally unchanged by a mismatch between blood flow and ventilation because increased elimination in overventilated areas compensates for decreased elimination in under ventilated areas. (We call this an imbalanced lung.) Consequently impaired CO_2 elimination has been considered to be diagnostic of some sort of blockage in the body's gas exchange system. J. W. Evans, P. D. Wagner, and J. B. West (1974) reexamined the question and found that Haldane was wrong: Unequal ventilation rates cause reduced CO_2 elimination. We develop a version of their model here.

Lungs function as follows. Air is drawn into the body, humidified, and pulled into little sacs in the lungs called alveoli. Here capillaries exchange CO_2 and oxygen (O_2) with the air, which is then exhaled and new air drawn in. If the blood flow around each alveolus were proportional to the volume of air in the alveolus, we would have a balanced lung. We want to compare CO_2 exchange in balanced and imbalanced lungs.

How much CO_2 is lost from the blood? At equilibrium the blood can hold a certain amount $C(P)$ of CO_2 per unit volume when the partial pressure, of CO_2 in the air is P. As CO_2 leaves the blood, P increases and the concentration in the blood decreases toward $C(P)$. For lack of better information, we assume that equilibrium is reached. Unfortunately P also increases as the blood absorbs oxygen, because P is proportional to the fraction of the air that is CO_2. It follows from the way carbohydrates and fats are used that over the long term the amount of CO_2 eliminated is about 80 % of the amount of O_2 taken in. If we assume that this is true for a single breath, we will have a constraint for the entire lung. This does not seem to be enough to give a manageable model; therefore we assume that this 80 % ratio holds for each alveolus for each breath. As you can see we are making a lot of unwarranted assumptions which may leave our conclusions on rather shaky ground. However, if after all these simplifying assumptions balanced and imbalanced lungs behave differently, it should be safe at least to conclude that Haldane was wrong.

Let's introduce some mathematical notation. We consider an individual alveolus first. Let the subscript i denote inspired and e denote expired. Let $P(x)$ denote the partial pressure of x. If we measure partial pressure in units so that atmospheric pressure is 1, then

(7) $$\sum_x P_i(x) = 1 \quad \text{and} \quad \sum_x P_e(x) = 1.$$

The change in the amount of x is (with suitable units)

$$(8) \qquad\qquad V_i P_i(x) - V_e P_e(x),$$

where V denotes the volume of air. Applying this to the gases,

$$(9a) \qquad\qquad CO_2 \text{ lost} = V_e P_e(CO_2) - V_i P_i(CO_2),$$

$$(9b) \qquad\qquad O_2 \text{ gained} = V_i P_i(O_2) - V_e P_e(O_2),$$

$$(9c) \qquad\qquad 0 = V_i P_i \text{ (other)} - V_e P_e \text{ (other)},$$

where the last equation is based on the fact that CO_2 and O_2 are the only gases exchanged in significant amounts. (Humidification occurs earlier.) We must combine (7) and (9) with the CO_2/O_2 ratio of 0.8 to obtain information about CO_2 exhaled, but in some simple form because we eventually will have to apply the result to all the alveoli and we can't measure individual volumes. Clearly the total volume change is 20% of the O_2 uptake. The CO_2 loss is 80% of the O_2 uptake, which is $(V_i - V_e)/0.2$ by the previous sentence. By (9a),

$$4(V_i - V_e) = V_e P_e(CO_2) - V_i P_i(CO_2).$$

Dropping the CO_2 in the P and rearranging,

$$V_i = \frac{V_e(4 + P_e)}{4 + P_i},$$

and so by (9a),

$$(10) \qquad\qquad CO_2 \text{ lost} = \frac{4V_e(P_e - P_i)}{4 + P_i}.$$

The object of all this is to compare balanced lungs with lungs in which air flow and blood flow are mismatched. Hence we need to supplement (10) with an equation involving blood flow. Let $C(P)$ be the concentration of CO_2 in the blood when the partial pressure of CO_2 in the air is P and equilibrium has been reached. Then for a quantity Q of blood passing by the alveolus and starting with a CO_2 concentration C_0,

$$(11) \qquad\qquad CO_2 \text{ lost} = Q[C_0 - C(P_e)],$$

if (a) the CO_2 balance in the air and blood reaches equilibrium before expiration and (b) the blood doesn't move (so that the blood coming by the alveolus at the start reaches the same CO_2 concentration as the blood coming by at the end). We've already decided to assume (a), but (b) doesn't look like a very good assumption. We should probably replace (11) by some sort of integral because blood is flowing by continuously. Since we can't handle this, we'll use (11) as an approximation, with Q equal to the quantity of blood flowing by in one breath.

We can now equate (10) and (11), the two expressions for CO_2 lost. The resulting equation can be solved for P_e, which can be substituted in (10) to obtain an expression for CO_2 lost which depends on V_e, Q, P_i, and C_0. The last two variables do not depend on the alveolus, and the first two enter only as a ratio except for a factor of V_e multiplying the entire expression. This sounds like a good approach, since changes in the ratio Q/V_e measure imbalance in the lung, and the V_e for the various alveoli add to a constant, the total volume of air exhaled. Let's carry out the plan. Let $g(x)$ be the solution to the equation

$$(12) \qquad \frac{4(g - P_i)}{4 + P_i} = x[C_0 - C(g)],$$

where we think of x as Q/V_e for applications. Letting the subscript a indicate a particular alveolus, the total CO_2 lost equals

$$\frac{\sum_a 4V_{ea}(P_{ea} - P_i)}{4 + P_i} = \frac{4[\sum_a V_{ea}g(Q_a/V_{ea}) - P_i \sum_a V_{ea}]}{4 + P_i}.$$

How does this change when total blood flow and total expired volume are held fixed? This is the question we must answer. Since P_i and $\sum V_{ea}$ are constants, it suffices to consider $\sum V_{ea}g(Q_a/V_{ea})$. For convenience, let's measure volume so that $\sum V_{ea} = 1$, and let's define the new variable $x_a = Q_a/V_{ea}$. In a balanced lung, x_a is constant. Hence showing that a balanced lung is more efficient at eliminating CO_2 is equivalent to showing that

$$(13) \qquad g\left(\sum_a V_{ea}x_a\right) > \sum_a V_{ea}g(x_a),$$

where the x_a are not all equal and the V_{ea} are positive numbers summing to 1. (You should convince yourself that this is what we need to do.)

Suppose a takes on only two values. Use Figure 2 to convince yourself that (13) holds if $g'' < 0$. Once this is done, it is fairly easy to prove inductively that $g'' < 0$ implies (13). We turn our attention to g''.

For the partial pressures associated with CO_2 in the lungs, $C(P)$ is nearly linear. Using this approximation, we can solve (12) for $g(x)$. [If solving (12) were impossible, we could use implicit differentiation to study g'' via (12).] Let $K = 4/(4 + P_i)$ and define A and B by $C(g) - C_0 = Ag - B$. Since C is an increasing function, $A > 0$. Equation (12) becomes

$$Kg - KP_i = Bx - Agx,$$

and so

$$g = \frac{KP_i + Bx}{K + Ax}.$$

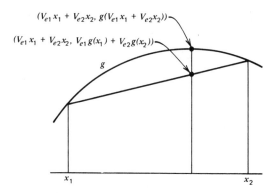

Figure 2   A graphical proof of (13) when a takes on two values.

Thus

$$g'' = -2AK(B - AP_i)(K + Ax)^{-3}.$$

By the definition of A and B, $B - AP_i = C_0 - C(P_i)$, which is positive since the blood gives up CO_2. Thus $g'' < 0$.

We've shown that imbalanced lungs have impaired CO_2 elimination, but this is based on some rather crude assumptions. Should we believe the result? First let's ask another question: Should we continue to accept Haldane's statement? Obviously not. In view of the present model it appears unlikely that his statement is correct, because it asserts that an equality holds—a very fragile prediction. However, inequalities are usually robust predictions. This by no means proves that our conclusion will stand up under improvement of the model, but it indicates that it is highly likely. I have looked at what I consider the two worst assumptions—the validity of (11) and the 80% ratio for each alveolus—but I don't see any reasonable way to improve them. Do you have any ideas?

PROBLEMS

1. This problem is based on H. M. Cundy (1971) and J. Higgins (1971). Suppose that you once owned a reel type tape recorder with a counter that counts revolutions of the take-up reel. Now you've replaced it with a recorder whose counter counts revolutions of the runoff reel. All your information concerning locations of songs on your tapes is now useless unless you can convert one counter value into the other.

Develop a method for doing this. Show how to construct a table for a given tape if you know the number of revolutions required to empty the reel and also the number of revolutions required to half empty the reel. (The half empty point is fairly easy to measure because the take-up and runoff reels will appear identical.) Do not assume that the thickness of the tape is known.

2. This problem was suggested by G. Levary (1956). A businessman is overstocked on a slow moving item. He wishes to mark down the price so that his overstock can be sold off to release money and space for other merchandise. What should he do? For uniformity we'll introduce the following notation:

L, list price of slow item.
L^*, proposed sale price.
S, number of slow moving items sold per year.
N, number of normal stock turnovers per year.
p, profit margin, that is, (net profits)/(total costs).

Consider the following questions and any others that come to mind. How many slow items should be retained? How low can the sale price be and still leave the merchant better off? If this problem is easy for you, here are some suggestions for complicating things. What if p can be higher on slow moving items because most people don't stock them? What about the effect of random fluctuations in demand? A Poisson model may provide a reasonable fit for the number of customers requesting a particular item during a time interval of some given length.

3. This problem is based on F. Metelli (1974). Certain mosaics of opaque colors give rise to the impression of transparency. We limit ourselves to shades of gray. With each shade one can associate a reflectance equal to the fraction of incoming light that is reflected. The range from black paper to white paper is about 4 to 80%. The left hand side of Figure 3 shows a mosaic made from four pieces with reflectances α_i. Under appropriate conditions it will appear to be two rectangular sheets which have been superimposed. The smaller sheet will appear to be semi-transparent, transmitting a fraction β of the incoming light. One necessary condition for apparent transparency is that the edge effects match up—discontinuities or even angles at a supposed boundary destroy the illusion of transparency. (Note that the central vertical line in Figure 3 is unbent where it crosses the boundary of the inner rectangle.) What conditions must the α_i, $i = 1, 2, 3, 4$, satisfy? How can we determine α_5 and β in terms of them? How would you test the model to see

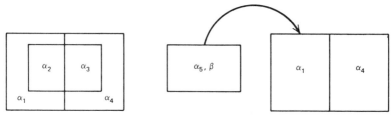

Figure 3   The mosaic on the left can be interpreted as the superposition of a semi-transparent sheet on a bicolored opaque sheet.

if the conditions are necessary? Sufficient? There are various interpretations for β which in turn lead to various formulas for α_2 and α_3. Consider

$$\alpha_2 = \qquad \alpha_5 + \beta\alpha_1,$$

$$\alpha_2 = (1 - \beta)\alpha_5 + \beta^2\alpha_1,$$

$$\alpha_2 = (1 - \beta)\alpha_5 + \beta^2\alpha_1[1 + (1 - \beta)\alpha_1 + (1 - \beta)^2\alpha_1 + \cdots]$$

$$= (1 - \beta)\alpha_5 + \frac{\beta^2\alpha_1}{1 - (1 - \beta)\alpha_1}$$

and any others that seem worth looking at. Which are correct?
Use it (or them) to answer the earlier questions.

4. Why do animals form herds? One obvious suggestion is protection against predators. What advantages does herding give to animals that always flee? Herding may reduce the chances of detection and capture *per prey* animal in the herd, and being near the middle of the herd may offer additional protection. Herding may also provide for improved detection of predators while grazing. Let's consider these by comparing a herd animal with a solitary animal in an open environment such as the African veldt. These ideas are adapted from I. Vine (1971) and H. R. Pulliam (1973). V. E. Brock and R. H. Riffenburgh (1959) discuss schooling of fish.

    (a) Let D be the distance at which a predator can be expected to detect a circular herd of n individuals and let d be the distance for a solitary animal. Argue that the chances of the herd being detected versus an isolated individual being detected are given by D^2/d^2 if the animals involved are placed at random on the veldt. Of course this doesn't happen; instead, the predator roams in search of prey. In this case can the relevant ratio be D/d? Explain?

(b) It is crucial to have an estimate for D/d. See Problem 1.5.6b. Show that $D/d \propto n^r$, with $0 \le r \le \frac{1}{2}$, may be a reasonable assumption. What can you say about r? What if predators detect prey by smell instead of sight?

(c) Suppose that animals mill around randomly within the herd. When is being in a herd safer than being isolated?

(d) In some herds, the animals push toward the center, with the result that some animals always end up on the perimeter. If a predator captures only animals that are on the perimeter, when is it safer to be on the perimeter of a herd than to be isolated?

(e) Criticize the following model and then develop it or an alternative model. A predator must get within some critical distance of a prey animal undetected in order to win the chase and make a kill; otherwise, the prey will escape. By looking up at random a grazing animal has some probability p of detecting the predator before it reaches the critical distance. Since one member of a herd can alarm the entire herd, a herd has a much better chance of escaping than an isolated individual. What is the probability that a herd will detect an approaching predator in time? There are some complications:

   (i) Not every herd member acts as a sentinel at the same time. (In some harems only the male performs sentinel duty, in some mixed herds some peripheral animals act as sentinels, etc.)

   (ii) If the predator approaches a large herd from a side opposite a sentinel, that sentinel won't spot the predator in time to alarm the herd.

(f) Taking (e) into account, return to (c) and (d).

(g) When herding is beneficial, what limits the size of herds? When is herding not beneficial? Can you add anything else to the subject of this problem?

5. The following is well known in traffic flow theory; see, for example, W. D. Ashton (1966, p. 18). Consider cars traveling along a roadway in one direction. Let k be the concentration of cars (e.g., the number of cars per 100 feet of roadway) and let q be the rate of flow (e.g., cars per minute).

(a) Argue that q and k are related as shown in Figure 4.

(b) Various implicit assumptions were needed in (a). State as many important ones as you can think of explicitly and defend and/or criticize them.

(c) Figure 4 is called a *fundamental diagram* or a *flow concentration curve*. Translate as many of the following as you can into traffic

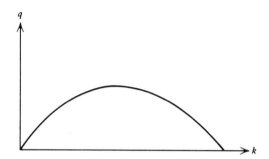

Figure 4 The fundamental diagram of traffic flow.

flow terms such as "speed on an empty roadway": (1) the values of
K and Q such that (K, Q) is the highest point on the curve; (2) the
slope of the line tangent to the curve at $(0, 0)$; (3) the slope of the line
tangent to the curve at (k, q); (4) the slope of the line connecting
$(0, 0)$ and a point (k, q) on the curve. *Hint*: If you don't know what
slopes measure, note that they have the same units as q/k.

(d) Does the above help organize and clarify traffic flow concepts for
you? What questions does it raise that may lead to further investiga-
tions and deeper understanding? In other words, what use is the
fundamental diagram?

6. When you view an object using only one eye, you can detect a change in
the brightness of the object if the change exceeds a certain threshold.
(See Problem 2.1.6.) Normally you use both eyes. Suppose we fool
the brain by exposing the eyes to separate but apparently identical
scenes whose brightnesses can be varied independently. A study of the
thresholds in this situation may give information about binocular vision.
This is what T. E. Cohn and D. J. Lasley (1976) did. They placed subjects
in front of a device that exposed the eyes as described above. The subject
reported pairs of left and right intensity changes (E_L, E_R) that resulted
in just noticeable changes in the apparently single object. Cohn and
Lasley plotted these points for various subjects and found that they lie
roughly on the ellipse $E_L^2 + E_R^2 + KE_L E_R = S^2$, where S depends on
the subject and $K \approx 0.6$. There is a fair amount of scatter in the data.
You will now consider various possible explanations for the data.

(a) Suppose that only the total intensity change matters. By "total" we
mean either $|E_L + E_R|$ or $|E_L| + |E_R|$. Describe the graphs Cohn
and Lasley could expect to obtain.

(b)   Suppose all that matters is that the change in at least one eye exceeds the threshold. Describe the graphs.

(c)   Combine the ideas in (a) and (b): It suffices to have the change in at least one eye ($|E_L|$ or $|E_R|$) or the change in both ($|E_L + E_R|$) exceed the threshold. Describe the graphs.

(d)   Cohn and Lasley proposed the following mechanism. The brain notes the sum and difference of E_L and E_R and combines them in some fashion to obtain a single parameter which must exceed a threshold. They suggest a weighted sum of squares: $(E_L + E_R)^2 + T(E_L - E_R)^2$. The value $T \approx \frac{1}{4}$ gives the ellipses mentioned earlier.

(e)   Compare the graphs in (c) and (d). They fit the published data about equally well. Where do we go from here? Can we decide between the models in (c) and (d) somehow, or decide that both are wrong?

PART 2

MORE ADVANCED
METHODS

CHAPTER 7

APPROACHES TO DIFFERENTIAL EQUATIONS

7.1. GENERAL DISCUSSION

Many phenomena can be described in a general way by saying that rates of change of the endogenous variables depend on past and present values of the variables. These situations lead to models involving differential and difference equations. The population models discussed in Section 1.4 are of this type: Equations (1) and (2) in Chapter 1 are differential equations, and (3) in Chapter 1 is a differential difference equation.

Models in the physical sciences frequently include force, which involves the second derivative of position with respect to time: $\mathbf{F} = d(m\, d\mathbf{x}/dt)/dt$, where $\mathbf{F}$ is force, m is mass, and $\mathbf{x}$ is position. The basic equations of electromagnetic theory are formulated in terms of partial differential equations. Thus the study of physical phenomena forces one to deal with differential equations.

Economics and sociology also deal with differential equations from time to time. See the marriage model in Problem 8.1.4. and the Keynesian model in Section 9.2 for examples.

Because of the importance of differential equations, the next two chapters are devoted to models involving ordinary differential equations. The rest of this chapter discusses some of the philosophy of studying differential equations and describes the topics covered and omitted in the next two chapters.

7.2. LIMITATIONS OF ANALYTICAL SOLUTIONS

It is usually best to solve the equations of a model exactly if the exact solution has a reasonable form. We call this an *analytical solution* of the model. If we find an analytical solution, we can often easily obtain information about the model that would otherwise be difficult or impossible to acquire.

The analytical approach has two severe limitations. The main one is that it may not be possible to solve the equations analytically, since the solutions of most equations cannot be found except numerically. Second, even if an analytical solution exists, it will not yield the desired information easily unless it is in a useful form. For example, it is not easy to see how $\sin x$ behaves for large values of x by considering the Taylor series expansion

$$\sin x = x - \frac{x^3}{6} + \frac{x^5}{120} - \cdots.$$

Nevertheless, analytical solutions are usually quite useful when they can be obtained. Models in this category are discussed in Section 8.1.

7.3. ALTERNATIVE APPROACHES

Since the analytical approach is often impossible or impractical, approximate methods are employed. These are roughly of two types: quantitative and qualitative. We usually put borderline cases in the latter category. What do we mean by these categories? Roughly speaking, "quantitative" refers to numbers and "qualitative" refers to shape, for example, "What is the value of $y(5)$?" versus "Is $y(t)$ periodic?" The following discussion should help to clarify this.

If you are interested in quantitative results, a computer is practically a necessity. The usual method for obtaining numerical information is to approximate the differential equations by difference equations and solve the latter. This sounds much easier than it is. We'd like a method that doesn't take a lot of computer time but gives a fairly accurate answer. We'd also like to know how accurate the computer's answer is. (It is important to remember that the computer's output is only an approximation. I know of one researcher who insisted on abandoning a model because the solution to his differential equation had small oscillations. They were present because of the method that was used in the computer center's differential equations package, but he insisted that the computer had *solved* his equation and that was that.) What we'd like and what we get may be two very different things. Very few computer centers provide differential equations packages that give error estimates, so you have to be a bit of a numerical analyst and try to obtain them yourself—if it can be done. We won't be concerned with numerical methods per se, but Section 8.2 contains some models for which numerical methods are useful.

In preliminary studies, when the data are very crude, or when the real situation is complicated, semiquantitative or qualitative statements are useful. Examples of such statements are

1. $f(t)e^{-2t}$ approaches a limit as $t \to \infty$.
2. For sufficiently large t, $x(t) > 0$.
3. (x, y, z) eventually approaches arbitrarily closely each point in D as $t \to \infty$.
4. $f(t)$ is bounded.

What are the advantages of such lack of precision over analytical and quantitative results? Because of the lack of precision, the model often need not be specified precisely. Thus we can often make robust statements about entire classes of models. This is useful in preliminary studies and in situations where the complexity precludes more accurate descriptions. Even if we have a specific model, we may wish to study the effect of certain parameters on the solution, for example, the effect of the amplitude and the length of the string on the period of a perfect pendulum:

$$(1) \qquad l\theta'' = -g \sin \theta, \qquad \theta(0) = A, \qquad \theta'(0) = 0.$$

In this case we can eliminate l and g by the change in variable $t = T(l/g)^{1/2}$ and solve the resulting equation. The period turns out to be given by what is known as an incomplete elliptic integral of the first kind:

$$(2) \qquad 2\left(\frac{2l}{g}\right)^{1/2} \int_0^A (\cos \theta - \cos A)^{-1/2} \, d\theta.$$

Since elliptic integrals have been studied extensively, quite a bit of information can be extracted from (2). Suppose we incorporate frictional effects by adding a term to the right hand side of (1) which depends on θ'. The analytical techniques collapse. If we know the precise form of the term that is being added to (1), we can conduct a time consuming numerical investigation. For a qualitative approach, see Section 9.2.

While some applications of qualitative methods to physics and biology are classic, the power of qualitative methods in modeling is just beginning to be realized. R. Thom's (1975) discussion of catastrophe theory has stirred up considerable interest.

7.4. TOPICS NOT DISCUSSED

Partial differential equations arise when we study variations of a function with regard to two or more parameters simultaneously. Except in the physical sciences, it is difficult to build models of this level of complexity without their becoming so complex that nothing can be done without a computer. Most exceptions seem to be based on physical analogies. Two very important partial differential equations are

Wave motion: $$\frac{a\,\partial^2 u}{\partial x^2} = \frac{\partial^2 u}{\partial t^2} \qquad a > 0.$$

Heat equation: $$\frac{a\,\partial^2 u}{\partial x^2} = \frac{\partial u}{\partial t} \qquad a > 0.$$

Equations like the first arise in the study of vibrating strings and membranes, and of electromagnetic, sound, and water waves. Equations like the second arise in the study of diffusion phenomena such as heat transfer, the spread of epidemics, and the change in gene frequencies in a population. Because sophisticated methods and/or extensive computer time are usually required to deal with partial differential equations, we avoid them.

Suppose we can relate the present state of a system to the state of the system at one or more previous times. The resulting equation is usually a difference equation. For example, suppose that female unicorns live for exactly 4 years and produce exactly one female offspring in their second and third years. Let $U(t)$ be the number of female unicorns at the end of year t. The number just born in year t is $U(t) - U(t-1)$, and they die in year $t + 4$ after bearing offspring in years $t + 2$ and $t + 3$. Thus

$$\begin{aligned} U(t) &= U(t-1) - [U(t-4) - U(t-5)] + [U(t-2) - U(t-3)] \\ &\quad + [U(t-3) - U(t-4)] \\ &= U(t-1) + U(t-2) - 2U(t-4) + U(t-5). \end{aligned}$$

This is an example of a linear constant coefficient difference equation. Models containing difference equations are designed to produce this type of equation because it is analytically tractable. Unfortunately they are often unrealistic. Attempts to add realism generally result in intractable equations which must be studied numerically. For these reasons as well as my own preferences, I've omitted difference equation models. The analytical intractability of difference equations is not wholly the result of neglect by mathematicians. Simple difference equations can have stranger solutions than simple differential equations, so that both analytical methods and qualitative methods are harder to develop. Although the differential equation $N' = rN(1 - N/K)$, where r and K are constants, has a very simple solution, the corresponding difference equation

$$N(t + 1) = N(t) + rN(t)\left(1 - \frac{N(t)}{K}\right)$$

is quite complicated. See R. M. May (1975). This richness of behavior may be useful in modeling when lots of computer time is available. However, it could prove embarassing—a model with too many possibilities is often worse than a model with too few.

In modeling populations the way we did unicorns, it is usually quite unrealistic to cut things up neatly into years. Attempts to avoid this often lead to integrals as a way of averaging over a period of time. Thus differential and difference equation models are closely related to integral equation models, another advanced topic that is not discussed here.

CHAPTER 8

QUANTITATIVE DIFFERENTIAL EQUATIONS

8.1. ANALYTICAL METHODS

In this section we consider models that lead to differential equations that have explicit solutions. With the partial exception of the ballistics model in Section 8.2, the examples were chosen to illustrate a variety of models, not to illustrate methods for solving differential equations.

Pollution of the Great Lakes

Industrialized nations are beginning to face the problems of water pollution. Once pollution of a river is stopped, the river will clean itself fairly rapidly if the pollution has not caused extreme damage. Lakes present a problem, because a polluted lake contains a considerable amount of water which must somehow be cleaned. The only presently feasible method is to rely on natural processes. How long does this take? In particular, how long would it take to clean up the Great Lakes?

Pollution affects a lake in many complex ways. Some compounds such as DDT enter biological systems and move up the food chain. Since DDT is very soluble in fat, it concentrates in the fatty tissue of higher predators and is hard to remove from the biosphere. Some pollutants move rather freely in and out of the food chain. The behavior of phosphorus lies somewhere between these two extremes. (In one sense, phosphorus is not a pollutant, since it occurs naturally; however, excessive amounts can trigger algae blooms, and it is then considered a pollutant.) Still other pollutants, like oil spills, may be only slightly involved in the food chain. Extensive pollution can cause irreversible damage and even "kill" a lake.

The main cleanup mechanism is the relatively straightforward natural process of gradually replacing the water in the lake. In addition, other processes such as sedimentation and decay may be important.

If we consider all these facets of the problem now, the discussion will go on and on and the resulting model will probably be hopelessly complex. Therefore we present the model first and discuss its validity later.

Figure 1 shows the Great Lakes. The numbers will be explained shortly.

The basic idea is to regard the flow in the Great Lakes as a standard perfect mixing problem. We ignore biological action, sedimentation, and so on, and assume that all the pollutants are simply dissolved in the water. This model is adapted from R. H. Rainey (1967).

We make the following assumptions:

1. Rainfall and evaporation balance each other, and so the average rates of inflow and outflow are equal.
2. These average rates do not vary much seasonally.

Figure 1   The Great Lakes. The figures indicate the number of years required to drain the lakes if outflow is unchanged and inflow stops.

These should be good approximations. In addition, we make the following rather questionable assumptions:

3. When water enters the lake, perfect mixing occurs, so that the pollutants are uniformly distributed.
4. Pollutants are not removed from the lake by decay, sedimentation, or any other mechanism except outflow.
5. Pollutants flow freely out of the lake—they are not retained the way DDT is.

By these assumptions, the net change in total pollutants during the time interval Δt is

$$\Delta(V P_l) = (P_i - P_l)(r \, \Delta t) + o(\Delta t),$$

where V is the volume of the lake, P_l is the pollution concentration in the lake, P_i is the pollution concentration in the inflow to the lake, r is the rate of flow, and $o(\Delta t)$ denotes a function of Δt such that $o(\Delta t)/\Delta t$ goes to zero as Δt goes to zero. Dividing this equation by Δt and letting Δt approach zero we obtain the differential equation

$$\tag{1} P_l' = \frac{(P_i - P_l)r}{V}.$$

Since this is a first order linear equation, we easily solve it to obtain

$$\tag{2} P_l(t) = e^{-t/\tau}\left[P_l(0) + \tau^{-1} \int_0^t P_i(x)e^{x/\tau} \, dx \right],$$

where $\tau = V/r$. The numbers in Figure 1 are Rainey's values of τ for the various lakes, measured in years. He does not give a value for Huron.

Using (2) and the data given in Figure 1 it is easy to determine the effect of various pollution abatement schemes *if the model is reasonable.* We do not include Lake Ontario in the discussion, because about 84% of its inflow comes from Erie, a source of pollution which can be controlled only indirectly. [The modifications required in (1) and the resulting time estimates are considered in Problem 1.]

The fastest possible cleanup will occur if all pollution inflow ceases. This means that $P_i = 0$. In this case (2) leads to the simple expression

$$\tag{3} t = \tau \log_e \left(\frac{P_l(0)}{P_l(t)} \right).$$

From this we can read off how long it would take to reduce pollution to a given percentage of its present level. The following figures in *years* were obtained in this fashion.

Lake	50%	20%	10%	5%
Erie	2	4	6	8
Michigan	21	50	71	92
Superior	131	304	435	566

Fortunately, the pollution in Superior is quite low at the present time.

We have built a very much simplified model. How much faith can we put in the times we have just obtained? To answer this question we must examine the validity of assumptions 3, 4, and 5.

We begin with the perfect mixing assumption. If a lake has only one source and one outlet, water tends to move from the source to the outlet in a pipeline fashion without mixing. Hence the cleanup time is shortened for the main part of the lake. (However, slow moving portions have much longer cleanup times.) This effect cannot push the times much below $V/r = \tau$, because a cleanup requires the replacement of nearly all the water in the lake. The value of τ is rather large for Michigan and Superior

Conclusion: Our assumption of perfect mixing may be far off, but this error is not likely to allow cleanup times much below τ and will probably lead to longer cleanup times for some semistagnant regions in the lake.

We discuss assumptions 4 and 5 in connection with two important pollutants: DDT and phosphorus. Mercury behaves like DDT in many ways, so the discussion applies to it as well.

Studies indicate that DDT and several other chlorinated hydrocarbons take a long time to break down into harmless compounds. Sufficient concentrations of DDT can have bad effects on the health of many organisms and even cause death. Unfortunately, DDT is almost impossible to remove from the biosphere. It dissolves readily in body fat, and so an organism retains most of the DDT in ingests. This causes the chemical to reach greater concentrations in higher predators. These animals are rather large and so are not likely to be swept out of the lake with the outflow unless they choose to leave. When an organism dies, most of its body fat is consumed by other organisms, so most of the DDT remains in the biosphere. As a result of all this, we can expect DDT to stay in the biota of a lake for an extended period of time. The main factor removing DDT from a lake may be its very slow breakdown into less noxious compounds, but consumption of fish by birds of prey and humans may be important.

Mercury behaves somewhat like DDT; however, it is an element and so does not decay. As a result, it is lost slowly due to sedimentation, outflow, and the removal of fish by birds and humans.

Phosphorus behaves differently. Large amounts of it are present in human wastes and in many fertilizers and detergents. The presence of excessive quantities of this element can cause algae blooms. These are sudden population explosions of algae as a result of which the lake may look like pea soup. Then the algae die and settle to the bottom. As a result, much of the phosphorus is removed in this fashion. Unfortunately, some of this removal is only temporary, since decay processes return the phosphorus to the lake water. The phosphate inflow to Lake Erie was about 75 tons daily in 1967, but the outflow was only about 25 tons (K. Sperry, 1967). Thus phosphorus was building up in the lake. The concentration may have been increasing, or the lake may have been losing 50 tons of phosphate per day in sediment. If the former is correct, cutting the inflow of phosphorus to 25 tons would only have led to an equilibrium situation. If the latter is correct, the phosphates on the bottom may reenter the biosphere and aggravate cleanup problems in the future.

Conclusion: For persistent pollutants like DDT the estimated cleanup times may well be too low. For other pollutants it is not clear how assumptions 4 and 5 affect the cleanup times.

Summary: The time estimates we derived may be low for some pollutants and high for others. The values of τ given in Figure 1 probably provide rough lower bounds for the cleanup times of persistent pollutants.

The Left Turn Squeeze

Have you ever found yourself in a car trapped near the curb with the rear end of a bus moving slowly and ominously toward you as the bus turns to the left? It can be a hair raising experience. How far to the right will the bus move? This model is adapted from J. Baylis (1973).

The situation is shown in Figure 2. We assume that the wheels do not slide sideways in turning. Since the rear axle is fixed, $\overline{FR}$ is tangent to the path of R. The angle between $\overline{FR}$ and the direction of the roadway is called φ, the length of $\overline{FR}$ is l, the length of $\overline{RT}$ is h, the width of the bus is $2w$, the turning angle of the front wheels is θ, and the speed of the bus is v. We must specify where the speed of the bus is measured. (To see this note that, if $\theta = 90°$ and the wheels don't slide sideways, the bus will move in a circle around R.) Let v be the speed of F. The values of φ, θ, and v are functions of time. Since we are interested only in the locus of U, we can take v to be any function of time. We set $v = 1$.

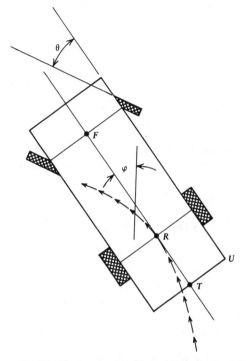

Figure 2 The bus turning left. Dotted line is path of R.

How can we describe the bus's motion? We sketch the derivation of the relevant equations, and you can fill in the details. By looking at the front end of the bus, we see that in a time interval dt the turning displaces the point F a distance

$$\sin \theta(v\, dt) = \sin \theta\, dt$$

perpendicular to $\overline{FR}$ and a distance $\cos \theta\, dt$ parallel to $\overline{FR}$. Looked at from the path of R, the displacement of F perpendicular to $\overline{FR}$ is $l\, d\varphi$, and the displacement parallel to $\overline{FR}$ depends on the path of R. Thus we have the basic equation relating φ, θ, and t:

(4) $$l\, d\varphi = \sin \theta\, dt.$$

We now turn our attention to the motion of U. Let us first compute the *leftward* displacement of F:

$$x(t) = \int_0^t \sin (\theta + \varphi) \, dt.$$

(We have used $v = 1$.) Using (4) and $\varphi = 0$ at $t = 0$, we obtain

(5)
$$x(\varphi) = l \int_0^\varphi \frac{\sin (\theta + \varphi)}{\sin \theta} \, d\varphi.$$

The displacement of U is now easily found: the *rightward* displacement of T is $(h + l) \sin \varphi - x$, and so the rightward displacement of U is

(6)
$$f(\varphi) = w \cos \varphi - w + (h + l) \sin \varphi - x.$$

Setting $f'(\varphi) = 0$, using (5), and multiplying by $\sin \varphi$, we obtain

(7)
$$[(h + l) \cos \varphi - w \sin \varphi] \sin \theta - l \sin (\theta + \varphi) = 0.$$

The general plan is to solve (7) for φ, use (5) to compute x, and then use (6) to compute the maximum displacement. To do this we need a relationship between θ and φ. Usually it is easiest if something is constant. Clearly φ cannot be constant, since the bus turns. Two possibilities are:

1. θ is constant—the driver keeps the front wheels turned at a constant angle relative to the bus.
2. $\theta + \varphi = \alpha$, a constant—the driver keeps the front wheels aimed in a constant direction relative to the roadway.

Possibility 1 is more realistic than possibility 2, but neither is perfectly correct. We consider both, because by comparing the results we should be able to obtain some idea of how accurate our conclusions are.

Suppose that θ is constant. Solving (7) and integrating (5) we obtain

(8a)
$$\cot \varphi = \frac{w + l \cot \theta}{h}$$

(8b)
$$x = \frac{l[\cos \theta - \cos (\theta + \varphi)]}{\sin \theta},$$

and the maximum displacement is

(8c)
$$f = [(w + l \cot \theta)^2 + h^2]^{1/2} - (w + l \cot \theta).$$

[The last equation is easily obtained by substituting (8b) into (6), expanding $\cos(\theta + \varphi)$, and recalling that the maximum of $A \cos \varphi + B \sin \varphi$ is $(A^2 + B^2)^{1/2}$; so we don't need (8a) for (8c).] Using (8) I obtained Table 1, based on the estimates $l = 16$, $h = 10$, and $w = 4$. The last row will be explained later.

Table 1   Maximum Displacement with θ Constant.

	θ					
	20°	30°	40°	50°	60°	70°
φ (degrees)	12°	18	23	30	37	46
f (feet)	1.0	1.5	2.1	2.7	3.4	4.2
$\bar{\alpha}$ (degrees)	26	39	52	65	79	93

Now let's consider the case in which $\theta + \varphi = \alpha$, a constant. Substituting $\theta = \alpha - \varphi$ into (7) we have, after rearranging,

$$[(h + l) \sin \alpha - w \cos \alpha] \cos 2\varphi - [(h + l) \cos \alpha + w \sin \alpha] \sin 2\varphi$$
$$+ (h - l) \sin \alpha + w \cos \alpha = 0.$$

Further rearranging gives

$$C \sin (2\varphi - \delta) = D,$$

where

$$C = [(h + l)^2 + w^2]^{1/2},$$
$$D = (h - l) \sin \alpha + w \cos \alpha,$$
$$\sin \delta = \frac{(h + l) \sin \alpha + w \cos \alpha}{C},$$

where $-90° < \delta < 90°$. Solving for φ,

(9a) $$\varphi = \frac{1}{2}\left[\arcsin\left(\frac{D}{C}\right) + \arcsin\left(\frac{E}{C}\right) \right],$$

where C and D are as before and

$$E = (h + l) \sin \alpha + w \cos \alpha.$$

Integrating (5),

(9b)
$$x = l \sin \alpha \log \left[\frac{\tan \alpha/2}{\tan (\alpha - \varphi)/2} \right].$$

Using (9) and (6) with $l = 16$, $h = 10$, and $w = 4$, as in Table 1, we obtain Table 2. The last row will be explained shortly.

How can we compare the two tables? After all, different things are constant in the two cases. A rough average value of α can be computed for Table 1 by noting that α varies between θ and $\theta + \varphi$ as φ varies between 0 and its optimum value. Thus we set $\bar{\alpha} = \theta + \varphi/2$. Likewise for Table 2, $\bar{\theta} = \alpha - \varphi/2$. Interpolating in Table 1 with the $\bar{\theta}$ of Table 2 used as θ, or

Table 2   Maximum Displacement with $\theta + \varphi$ Constant.

	α					
	20°	30°	40°	50°	60°	70°
φ (degrees)	7	11	15	18	22	26
f (feet)	0.8	1.1	1.5	1.9	2.3	2.7
$\bar{\theta}$ (degrees)	16	24	33	41	49	57

doing the similar thing with the tables interchanged and using $\bar{\alpha}$ instead of $\bar{\theta}$, we see that the estimates of f are within about 20% of each other. This suggests that a table of $\bar{\theta}$ (or $\bar{\alpha}$) versus the maximum f will be about the same for almost any method of turning. How could you test this idea? Thus we conclude that the rear end of a bus turning left moves about $1\frac{1}{2}$ feet to the right, or more if the driver makes a sharp turn.

Long Chain Polymers

Our booming synthetic fabric industry relies on chemical reactions that produce long chain organic polymers. Thus it is important to understand the nature, speed, and end products of polymerization reactions. We study one type of reaction here and another in Problem 3. The material is adapted from C. Tanford (1961, Ch. 9).

We need some background in chemistry. A simple reaction of the type we wish to study is

$$
\begin{array}{c}
\text{O} \\
\parallel \\
\text{RCH} - \text{C} \\
\quad\diagdown \\
| \qquad\quad \text{O} \; + \; \text{H} - (\text{NH} - \text{RCH} - \text{CO})_n - \text{R}' \\
\quad\diagup \\
\text{NH} - \text{C} \\
\qquad\diagdown \\
\qquad\quad \text{O} \qquad\longrightarrow\qquad \text{H} - (\text{NH} - \text{RCH} - \text{CO})_{n+1} - \text{R}' + \text{CO}_2 ,
\end{array}
$$

where $n \geq 0$ and the radical R' provides the mechanism for the reaction by breaking open the anhydride ring. We write the reaction symbolically:

$$(10) \qquad\qquad A + M_n \longrightarrow M_{n+1} + CO_2 .$$

The compound M_n is called a *polymer of length n*. For fixed temperature and pressure, the rate of a chemical reaction like (10) depends on the probability of a collision between an A molecule and an M_n molecule. This is proportional to the product of their concentrations, which is written $[A][M_n]$. Thus the *rate* of reaction (10) is $k_n[A][M_n]$, where the *rate constant* k_n is practically the same for all n because the reaction mechanism is the same. We assume $k_n = k$ for all n. So much for background.

A typical process starts with a concentration $a(0)$ of A and a concentration $m_0(0)$ of M_0 (which is simply $R'H$). How does the system evolve? To begin with, since the concentration of R' does not change, we have the conservation equation

$$(11) \qquad\qquad \sum_{n=0}^{\infty} m_n(t) = m_0(0),$$

where $m_n(t)$ is $[M_n]$ at time t. From (10) we have

$$(12a) \qquad\qquad \frac{dm_0}{dt} = -ka(t)m_0(t),$$

$$(12b) \qquad\qquad \frac{dm_n}{dt} = ka(t)[m_{n-1}(t) - m_n(t)], \qquad n \geq 1,$$

$$(12c) \qquad\qquad \frac{da(t)}{dt} = -ka(t)\sum_{n=0}^{\infty} m_n(t).$$

Combining (11) and (12c), we obtain

$$\frac{da(t)}{dt} = -km_0(0)a(t),$$

which has the solution

(13a) $$a(t) = a(0)e^{-\lambda t}, \qquad \lambda = km_0(0).$$

We can simplify (12a) and (12b) by defining a variable y such that

(13b) $$dy = ka(t)\,dt, \qquad y = 0 \text{ at } t = 0,$$

for (12a) and (12b) can then be rewritten as

(14a) $$\frac{dm_0(y)}{dy} = -m_0(y),$$

(14b) $$\frac{dm_n(y)}{dy} = m_{n-1}(y) - m_n(y).$$

These equations are easily solved inductively to obtain

$$\frac{m_n(y)}{m_0(0)} = \frac{e^{-y}y^n}{n!},$$

a Poisson distribution with parameter y. (Do it.) Thus the mean chain length is y, and the variance of the length is also y. We now use (13) to determine y as a function of t:

(15) $$y(t) = \int_0^t ka(0)e^{-\lambda t}\,dt = \frac{a(0)}{m_0(0)}(1 - e^{-\lambda t}).$$

How can we produce polymers of some desired length l? Setting $y = l$ we obtain

(16) $$t = \frac{-\log\left[1 - m_0(0)l/a(0)\right]}{km_0(0)}.$$

Since the Poisson distribution can be approximated by a normal distribution when y is large, about 95% of the lengths lie between $l - \sqrt{l}$ and $l + \sqrt{l}$. Note that altering reaction conditions like temperature and pressure only affects the time t that we let the reaction run and has no effect on the distribution of final chain lengths.

Let's examine briefly what happens if we relax the assumption that $k_n = k$. If k_n is a decreasing function of n, the reaction proceeds more slowly than expected as time goes on, because the polymers are becoming longer. Also, the final distribution of chain lengths is more peaked than a Poisson distribution, because the shorter chains increase in length faster than the longer chains. You should be able to explain what happens when k_n is an increasing function of n.

How can we use these results in chemical engineering? We can use (16) to determine the optimum values for $m_0(0)$ and $a(0)$. To make the reaction run as fast as possible, both $m_0(0)$ and $a(0)/m_0(0)$ should be large. Since there is an upper limit to the possible combined concentrations of A and M_0—only so much will fit in a given volume—we obtain an inequality:

(17) $$0 \leq a(0) \leq f(m_0(0)),$$

where $f' < 0$. (Why?) As already noted, the larger $r = a(0)/m_0(0)$ is, the faster the reaction proceeds. Since fast reactions save time, increasing r increases the number of batches we can process. Unfortunately, when we stop the reaction the concentration of A remaining will be

(18) $$\alpha = m_0(0)(r - l).$$

Thus, if we cannot reclaim the remaining A or if the reclamation expense increases with quantity, a larger r will increase our expenses. By studying the details of plant operation we can construct a cost function depending on $t, \alpha, m_0(0)$, and r, where t and α are given by (16) and (18). We can then minimize this subject to (17). In this way it is possible to reduce costs considerably over what they might be for a naive approach to plant design.

PROBLEMS

1. This problem relates to the pollution of Lake Ontario.

   (a) Use the subscript e to refer to Erie, the subscript o to refer to Ontario, and the subscript i to refer to non-Erie inflow to Ontario. Show that (1) should be replaced by

   $$P_o' = \frac{P_e r_e}{V_o} + \frac{P_i(r_o - r_e)}{V_o} - \frac{P_o r_o}{V_o}.$$

   (b) Using the fact that about five-sixths of the inflow of Ontario is the outflow from Erie, deduce that

   $$P_o(t) = e^{-t/\tau}\left\{P_o(0) + \frac{1}{6\tau}\int_0^t [5P_e(x) + P_i(x)]e^{x/\tau}\,dx\right\}.$$

   (c) Assuming that all pollution inflow to Erie and Ontario ceases except for the uncontrollable flow from Erie to Ontario, compute the 50 and 5% cleanup times for Ontario. To do this, you will need to know how the pollution level of Erie compares with that of Ontario. No data are available on this, but Erie seems to be more polluted. Try various values for $P_e(0)/P_o(0)$.

(d) In the model we discussed the effect of various types of pollutant behavior on cleanup times. If necessary, reconsider this for Ontario.

2. This problem deals with simple *compartment* models in physiology. See D. S. Riggs (1963) for further discussion, especially his Sec. 6-14 which treats problems of fitting curves to such models.

   (a) Treat the blood as a compartment containing a substance being removed by a physiological mechanism. What sort of equations could describe the concentration of the substance as a function of time? We need *simple* models. How can they be tested?

   (b) Let's be specific and assume that the removal is being done by the kidneys. In this case the rate of removal is usually proportional to the amount of the substance passing through a kidney per unit time. Construct a simple model based on concentrations.

   (c) The substance in (b) is a drug whose concentration should lie between 2 and 5 milligrams per 100 cubic centimeters. If the drug is taken internally, about 60% is quickly absorbed and most of the remainder is lost. In about 8 hours the body of an average person eliminates about 50% of the drug. A normal adult has about 5 liters of blood. Design a dosage program for the drug.

   (d) Most drugs are taken orally and require time to be absorbed by the blood. At the same time the drug is being removed by the kidneys. Model the situation. Here is some data on drugs taken from J. V. Swintosky (1956). The first drug is sulfapyridine, and the second is sodium salicylate. An O indicates oral administration, and an I indicates intravenous administration [to which (a) should apply]. The column headed "grams" gives the initial dosage, and the other columns indicate the concentration in the blood at various times after administration. How well does your model fit? Could you explain any discrepancies?

Concentration (milligrams/cubic centimeters)

Administration	Grams	1 hour	2 hours	4 hours	6 hours	8 hours	10 hours	12 hours	24 hours
O	4.0	2.3	2.7	3.6	3.0	—	2.0	—	—
O	4.0	1.8	2.8	3.9	3.5	2.6	2.2	—	—
I	1.8	3.8	3.4	2.6	2.1	—	—	—	—
I	1.8	3.7	3.3	2.7	2.3	—	—	—	—
O	10	5.0	—	—	14.4	—	—	15.7	12.5
I	10	39.4	—	—	31.4	—	—	24.2	16.2
I	20	56.7	—	—	43.0	—	—	35.2	26.6

For a further discussion of drug kinetics see R. E. Notari (1971).

(e) General anesthetics are usually administered through the lungs. What factors do you think are important in modeling anesthetic concentration in the blood? Outline a model. The rate of absorption through the lungs may vary considerably from one substance to another. An anesthetist monitors an anesthetized patient to decide how to adjust the flow of anesthetic. Do you think the absorption rate should be taken into account? Explain.

3. Another sort of polymerization reaction is called *condensation*. A simple reaction of this sort is

$$M_k + M_n \longrightarrow M_{k+n} + H_2O.$$

Study $[M_n(t)]$. *Warning*: Counting reactions is a bit tricky; don't count $M_k + M_n$ and $M_n + M_k$. Also, beware of $M_n + M_n$, because $[M_n]^2$ counts each collision twice.

4. At what age are your friends going to be marrying most rapidly? 15? 20? 25? 30? What factors cause people to marry? Sociologists and psychologists generally believe that peer group behavior plays a major role. Can we model this? The following attempt is adapted from G. Hernes (1972).

(a) It is assumed that a person's chances of marrying in some small time interval Δt are proportional to Δt and to the fraction of people in the person's age group that are already married $m(t)$. This is based on the idea that there is overt and covert peer group pressure to marry. Show that this leads to the differential equation

$$m' = cm(1 - m).$$

Solve the equation.

(b) The model may be criticized for a variety of reasons; for example, it assumes that all people feel the same pressure to marry regardless of individual and age as long as the fraction of the peer group that is married is the same. Discuss the model critically.

(c) Suppose $c = c(t)$. How can this help the model? What is the solution to the differential equation? In terms of properties of $c(t)$, determine what fraction of people in your age class will eventually marry.

(d) Hernes finds that

$$\log [c(t)] = ab^t \log k, \qquad b < 1,$$

gives a rather good fit, but a variety of other forms for $c(t)$ may do just as well. Can you suggest properties a good $c(t)$ is likely to have?

(e) We have ignored the problem caused by the fact that, since $m(t)$ was zero when your peer group was younger, the differential equation predicts that it will remain zero. How can we get around this? Remember that we are trying to provide a model that will roughly fit the situation.

(f) Discuss how to handle the fact that people are not identical. Can this be incorporated in $c(t)$ somehow? (We could expect the average value of c to decrease with time as those who are more likely to marry do so.)

(g) A. J. Coale (1971) found that, by making a linear transformation of the age axis, $x = at - b$, and a scale transformation of the proportion married axis, $y = m/m(\infty)$, a curve was obtained that was closely fitted by

$$y = \exp(-e^{-x})$$

How does this fit in with the previous discussion? (Coale used data from a variety of countries; Hernes used data from a U.S. census.) K. C. Land (1971) discusses a Poisson model for divorce.

5. How long does it take an object to fall from a great height? You may need some or all of the following facts:

1. The drag force on similarly shaped objects depends on the density of the air ρ, the velocity of the object v, the speed of sound c, and a characteristic dimension of the object d.
2. The velocity of sound c depends on the pressure p and density ρ of the air.
3. If h is the height above the ground, $dp = -gp\,dh$, where g is acceleration due to gravity.
4. Pressure satisfies $p \propto \rho T$, where T is temperature in degrees Kelvin.
5. The force of gravity is mg, where $g \propto r^{-2}$ and r is the distance from the object to the center of the earth. The radius of the earth is about 4000 miles.

Before plunging in blindly and trying to build a model that uses all these facts, you had better consider just what it is you want to know. The problem is rather vague: How great a height? How accurate an answer? Of course you may decide you need all these facts and some additional ones besides. Whatever you decide, come up with a reasonable method for obtaining an answer of some sort.

6. What is the best way for our company to run its advertising campaign? A variety of models has been developed to study the effects of advertising on consumer behavior by people who do marketing research. The more elaborate models often allow for more than one type of consumer behavior, each type having at least two and sometimes several constants to estimate. Obviously one can fit data better with complicated models, but frequently one such complicated model is about as good as another. This is a delicate, data-hungry approach. Here you should develop the simplest model you can.

It has been observed that in our company's markets consumer purchases drop off roughly like exponential decay when advertising stops. (This is often a fairly good approximation in real life.) It seems reasonable to assume that new customers are attracted by advertising at a rate that depends on the fraction of the potential market that does not buy our product and on the level of our advertising.

(a) Construct a simple differential equation model based on these ideas. Criticize it.

(b) Make some predictions that could be used as tests for your model. (Remember the expense that may be involved.)

(c) How should our company spend its advertising budget for the next 6 months—on an intensive 2 week campaign with little additional advertising, or on a uniform advertising plan for the entire 6 months? You have to make and defend a recommendation as a part of your job. Do you need additional data which the company can provide for you? How much faith do you have in your suggestions?

(d) What should our total advertising budget be? How much should we spend on market research? Why? (Remember, it's your job you're discussing.)

M. L. Vidale and H. B. Wolfe (1957) discuss some of these problems.

7. J. S. Coleman (1964, Sec. 8.4) discusses a model of the effect of an insecticide on the death rate of insects. He makes the simplifying assumptions:

1. Above a certain threshold an increase Δc in insecticide concentration causes a fraction $\alpha \Delta c$ of those insects that would have survived the old concentration to die.

2. Below the threshold the insecticide has no effect.

3. For most insecticides we can expect an additive effect; that is, if the parameters for two insecticides are α_1 and α_2 and if their fractions in a mixture are p and $1 - p$, the combination has the parameter $\alpha = p\alpha_1 + (1 - p)\alpha_2$; and the thresholds combine similarly.

He presented the following data (due to Finney) on the fraction of houseflies killed using rotenone and pyrethrins in various proportions. Here c is the concentration in milligrams per cubic centimeter and d is the fraction dying. Two series were run on the unmixed pesticides.

Rotenone			Pyrethrins			1:5 Mixture		1:15 Mixture	
c	d	d	c	d	d	c	d	c	d
0.10	0.24	0.28	0.50	0.20	0.23	0.30	0.27	0.40	0.23
0.15	0.44	0.51	0.75	0.35	0.44	0.45	0.53	0.60	0.48
0.20	0.63	0.72	1.00	0.53	0.55	0.60	0.64	0.80	0.61
0.25	0.81	0.82	1.50	0.80	0.72	0.875	0.82	1.20	0.76
0.35	0.90	0.89	2.00	0.88	0.90	1.175	0.93	1.60	0.93

(a) Develop Coleman's model and test it against the data.
(b) Assuming rotenone and pyrethrins act independently (which Coleman's model translates as "additively"), can you develop other simple models with some reasonable notion of independence that fit the data as well as Coleman's model?

8.2. NUMERICAL METHODS

In this section we are not concerned with how the actual numerical solution of a problem is carried out, but rather with models that lead to a need for numerical solutions. A variety of numerical methods exists in the literature, and most computing centers have at least one package for solving differential equations numerically. If you wish or need to write your own package, a simple numerical technique is given at the end of this chapter.

Towing a Water Skier

You may have noticed that a water skier tends to slow down when the boat towing him turns. Two factors influence this: (1) For the same amount of power, a turning boat travels slower than a boat moving on a straight course, and (2) the skier tends to follow a shorter path than the boat. Can we model the situation?

Let's look at the skier first. If he drops the tow rope, he will lose speed very rapidly because of the drag of the water. Thus the skier always moves practically along the line of the tow rope unless he can do something to affect his direction of motion. He can exert some control through the position in which he holds his skis in the water. To avoid this rather grave complication, we assume that the skier does the easiest thing and keeps his skis pointed toward the boat. Thus we have created a skier whose rope is always taut (because of the drag of the water) and who always moves in the direction of the rope. Let the tow rope length be l, the coordinates of the rear of the boat be $[x(t), y(t)]$, and the coordinates of the skier be $[r(t), s(t)]$. By considering the length of the rope and the direction of motion of the skier we obtain two separate equations:

(19a)
$$l^2 = (r - x)^2 + (s - y)^2,$$

(19b)
$$\frac{s'(t)}{r'(t)} = \frac{s - y}{r - x}.$$

We manipulate these two equations to obtain a set of two first order equations for $r(t)$ and $s(t)$. By differentiating (19a) with respect to t, clearing fractions in (19b), and rearranging each of them, we obtain two equations in r' and s':

(20)
$$2(r - x)r' + 2(s - y)s' = 2(r - x)x' + 2(s - y)y',$$
$$(s - y)r' + (r - x)s' = 0.$$

Solving for r' and s' and using (19a), we obtain

(21)
$$r' = \frac{x'(r - x)^2 + y'(r - x)(s - y)}{l^2},$$
$$s' = \frac{y'(s - y)^2 + x'(r - x)(s - y)}{l^2}.$$

Before we can solve (21) we must model the motion of the boat. Knowing the boat's course is enough to let us determine the skier's course: If we know y as a function of x, multiplying (21) by dt/dx gives a set of two differential equations which can be solved numerically for r and s as functions of x. This gives us the path of the skier parametrically in terms of x. We now determine his speed in terms of the boat's speed v. The x component of the boat's velocity is $v/[1 + y'(x)^2]^{1/2}$ by basic calculus and geometry. Thus the skier's speed is

$$\sqrt{r'(x)^2 + s'(x)^2}\,x'(t) = \sqrt{r'(x)^2 + s'(x)^2}\,x'(t) = v\sqrt{r'(x)^2 + s'(x)^2}/[1 + y'(x)^2].$$

Hence the skier's speed at any time equals the boat's speed at that time multiplied by some function of the paths of the skier and the boat. Alternatively, we can solve (21) under the assumption that the boat's speed always equals 1. We will obtain the path of the skier and, by the argument just given, a "speed" for the skier which is equal to the skier's true speed divided by the boat's true speed. This enables us to treat the problem of the boat's speed as a completely separate issue. Since it is a complicated hydrodynamic problem, we do not attempt to solve it. Consequently we obtain only a partial solution to the problem we started out with; however, the full solution will be easy to find if and when we obtain information on the speed of a speedboat making a turn.

We could try all sorts of paths for a turn. The simplest to program is a circular arc, and this is a reasonable path. By defining

$$(22) \qquad x(t) = Bl\cos\left(\frac{t}{Bl}\right) \quad \text{and} \quad y(t) = Bl\cos\left(\frac{t}{Bl}\right),$$

I obtained a circular course with radius equal to B rope lengths and a speed of 1. I decided it would be interesting to note how far the angle of the rope deviated from a line straight back from the boat. By substituting (22) into (21) and integrating numerically I found that with $B = 1$, a *very* sharp turn, the speed of the skier dropped markedly: After a 90° turn by the boat his speed was 67% of the boat's and his angle with the line of the boat was 47°. After a full 180° turn the figures were 45% and 63°. By the time the radius of the turn was twice the tow rope length the situation had improved considerably: The skier's speed was still 86% of the boat's speed after a 180° turn, and his angle was only 30°. The changes were fastest at the start of the turn; in fact, after 45° the skier's speed had already dropped to 92%, and his angle was 23°. With a turn of radius four times the tow rope length the speed change was negligible—still 96% of the boat's speed after 180°. The tow rope's angle with the line of the boat was only 14°. The lesson is quite clear: To keep up a water skier's speed be sure the radius of your turn is at least twice the tow rope length. A radius four or more times the tow rope length results in almost no loss in the speed of the skier except for a possible loss due to the boat slowing in the turn. Alternatively, the skier can maintain his speed by pointing his skis somewhat outward from the direction of the turn so that he does not move in the direction of the rope. The analysis of this situation appears complicated.

We seem to have completed the problem. This was my reaction until I examined the data a bit more closely.

θ	B = 1		B = 2		B = 4	
	φ	w	φ	w	φ	w
0°	0°	1.00	0°	1.00	0°	1.00
15°	13°	0.97	12°	0.97	9°	0.98
30°	23°	0.91	19°	0.94	13°	0.97
45°	31°	0.85	23°	0.92	14°	0.97
60°	38°	0.78	26°	0.90	14°	0.96
75°	43°	0.72	27°	0.88	14°	0.96
90°	47°	0.67	28°	0.88	14°	0.96
105°	51°	0.62	29°	0.87	14°	0.96
120°	54°	0.58	29°	0.87	14°	0.96
135°	57°	0.54	30°	0.87	14°	0.96
150°	59°	0.51	30°	0.86	14°	0.96
165°	61°	0.48	30°	0.86	14°	0.96
180°	63°	0.45	30°	0.86	14°	0.96

It is reproduced here. The angle the boat has turned through is θ, the water skier's angle with the boat is φ, and his speed divided by the boat's is w. (Incidentally, finding the formula for φ is a nontrivial problem. You should do it.) Note that w appears to depend only on φ. Let's prove this for any motion and compute the function $w(\varphi)$. For simplicity we move the coordinate system so that at $t = 0$ the boat is at the origin and its direction of motion is along the x axis. Hence we have at $t = 0$,

$$x = y = 0, \qquad x' = 1, \qquad y' = 0, \qquad \cos \varphi = \frac{-r}{l},$$

$$w^2 = \left(\frac{dr}{dt}\right)^2 + \left(\frac{ds}{dt}\right)^2.$$

By (21) $r' = r^2/l^2$ and $s' = rs/l^2$. Hence

$$w^2 = \frac{r^4 + r^2 s^2}{l^4} = \frac{r^2}{l^2},$$

and so $w = \cos \varphi$. Such a simple formula is unlikely to depend on more than a simple geometric argument. Can you find one?

A Ballistics Problem

During World War II, mathematicians were asked to construct tables for gunners relating angle to range. Bombadiers required similar information. How was this done? In this case the model is fairly straightforward, and the emphasis is on the mathematics, in contrast to most other models we have studied.

We wish to construct a model of the' motion of an object under the influence of gravity and air resistance. This material is adapted from T. v. Kármán and M. A. Biot (1940, pp. 139–143). We ignore the complications due to lifting forces and possible rotation of the object. Hence the only forces involved are a downward force of mg and a drag force opposite the direction of motion of $mf(v)$, where m is the mass of the object, $v = |\mathbf{v}|$ is the magnitude of its velocity, and g is the acceleration due to gravity. In an $x - y$ coordinate system with the positive y axis directed downward, we can write this as a vector equation: $\mathbf{v}' = (0, g) - [f(v)/v]\mathbf{v}$. Over a fairly large practical range, $f(v)$ is nearly proportional to v^2.

We let θ be the angle between $\mathbf{v}$ and the x axis and resolve the acceleration into components parallel and perpendicular to $\mathbf{v}$. To do this we need to know the value of $\mathbf{v}'$ in the two directions. Since

$$\mathbf{v} = (v \cos \theta, v \sin \theta),$$

$$\mathbf{v}' = (\cos \theta, \sin \theta)v' + (-v \sin \theta, v \cos \theta)\theta',$$

the parallel component is simply v' and the perpendicular component is $v\theta'$. Resolving the acceleration due to gravity into components parallel and perpendicular to $\mathbf{v}$ and using the fact that drag acts parallel to $\mathbf{v}$, we obtain

(23a) $$v' = g \sin \theta - f(v),$$

(23b) $$v\theta' = g \cos \theta.$$

Multiplying (23a) by $vg \cos \theta$, dividing by (23b), and rearranging, we obtain

$$\frac{g\, d(v \cos \theta)}{d\theta} = -vf(v),$$

an equation we cannot solve analytically unless f has some special form. If we assume that $f(v) = kv^2$, we obtain

$$g\, dv_x/v_x^3 = -\frac{k\, d\theta}{\cos^3 \theta},$$

where $v_x = v \cos \theta$, the component of $\mathbf{v}$ in the x direction. Hence

$$v_x^{-2} = \frac{2k}{g} \int \cos^{-3} \theta\, d\theta.$$

Suppose that $\mathbf{v} = (v_0, 0)$ when $\theta = 0$. Carrying out the integration and rearranging,

$$(24a) \qquad v_x = r(\theta), \qquad v_y = r(\theta) \tan \theta, \qquad v = \frac{r(\theta)}{\cos \theta},$$

where

$$(24b) \qquad r(\theta) = v_0 \left[1 + \frac{k v_0^2}{g} \left(\frac{\sin \theta}{\cos^2 \theta} + \log \frac{1 + \sin \theta}{\cos \theta} \right) \right]^{-1/2}.$$

We now integrate these velocity equations to obtain the path of the object. Let the origin be at $\theta = 0$. Using in succession the chain rule, $v_x = v \cos \theta$, and (23b), we have

$$(25) \qquad \frac{dx}{d\theta} = \frac{v_x}{\theta'} = \frac{v \cos \theta}{g \cos \theta / v} = \frac{v^2}{g}.$$

Similarly,

$$(26) \qquad \frac{dy}{d\theta} = \frac{v_y}{\theta'} = \frac{v \sin \theta}{g \cos \theta / v} = \frac{v^2 \tan \theta}{g}.$$

Combining (24a) with (25) and (26) we obtain the coordinates parametrically in terms of θ:

$$(27) \qquad [x(\theta), y(\theta)] = \left[\int_0^\theta \frac{r(\theta)^2 \, d\theta}{g \cos^2 \theta}, \int_0^\theta \frac{r(\theta)^2 \sin \theta \, d\theta}{g \cos^3 \theta} \right].$$

Since $r(\theta)$ is given by (24b), the integrations in (27) can be carried out numerically. And alternative approach is to solve the original differential equations directly by numerical methods. This is more sensitive to numerical errors, because the original equations are linked second order equations while (27) simply involves two disjoint integrals.

We can supplement (27) by obtaining time information. Using (24a) to eliminate v in (23b), $r(\theta)\theta' = g \cos^2 \theta$. Thus

$$(28) \qquad t = \int_0^\theta \frac{r(\theta) d\theta}{g \cos^2 \theta}.$$

Since our time origin is at $\theta = 0$, we must integrate back from 0 in (27) and (28) to obtain the intial position for a projectile fired upward.

PROBLEMS

1. Consider the left turn squeeze model in Section 8.1.

   (a) Discuss in class how you could take steps toward answering the question raised at the end of the model by using a computer: How can we show that the results are not very sensitive to the form of $\theta(t)$ for reasonable methods of turning? (Or perhaps, discover that they are.) Be specific.

(b) If the class has access to a computer, implement the plan formulated in (a).

2. In this problem the question is: How can we formulate a model that does not require an excessive amount of computer time? Most galaxies appear to be fairly flat disks with the stars moving about a common center like a huge swarm of planets or asteroids. Nearly all the mass of the galaxy is in the central region, because the stars there are much closer together. Some astronomical photographs (A. Toomre and J. Toomre, 1973) show pairs of galaxies which appear to have collided, or at least passed close to one another and caused large streamers of stars to be pulled out. How could you test this idea using a mathematical model? Recall Newton's law of gravity is $F = Gm_1m_2/r^2$, directed along the line between two bodies, and Newton's basic law is $F = ma$. See A. Toomre and J. Toomre (1972, 1973) afterward if you want to see how they did it.

3. This problem is adapted from M. S. Bartlett (1972). Can we construct a simple model of the spread of epidemics? We take as our example measles, a prevalent childhood disease before vaccinations became available. The incubation time is $\frac{1}{2}$ week. During this time a child seems normal but is able to infect others. After this time the child is isolated until recovery, at which point he or she is immune. Roughly speaking, measles outbreaks have been more severe during alternate years.

(a) Construct a simple differential equation model allowing for three categories: susceptible, infective, and isolated/recovered. Allow for an influx of new susceptibles due to births. Assume an infective makes contact with members of the population at random and infects a contacted susceptible with probability p.

(b) Construct a simple difference equation model.

In what follows use the differential equation model, the difference equation model, or both.

(c) Show that your model has some sort of cyclic behavior. If it doesn't, fix it, because measles outbreaks definitely tend to occur in a cyclic pattern.

(d) Estimate the parameters in your model to fit the $\frac{1}{2}$ week incubation and 2 year cycle observations. Do the parameter values appear realistic?

(e) Measles outbreaks are seasonal (60% below average in summer and 60% above average in winter), but if you've constructed a model of the sort I expected, a slight change in the parameters in (d) will cause the period to differ slightly from 2 years and so the peak will drift from season to season. What can be done? Most children make contact with more children during the school year than during vacation. Use this to fix up the model by introducing a seasonal variation in p. How much variation is required? Does this amount seem reasonable?

(f) Can you allow for contact between school districts?

(g) How much faith do you have in the model? What are its faults? Can you suggest improvements? The following data from Bartlett's article may be useful.

Annual measles deaths in London (1647–1660)

1647	1648	1649	1650	1651	1652	1653
5	92	3	33	33	62	8

1654	1755	1656	1657	1658	1659	1660
52	11	153	15	80	6	74

Mean time between epidemics for some towns
in England and Wales (1940–1956)

Population (thousands)	Time between outbreaks (weeks)
1046	73
658	106
415	92
269	93
180	94
113	80
66	74
22	86
18	92
12	79
11	98
7	199
4	105

4. Organisms have internal oscillations, like circadian rhythm, which have natural periods, like 24 hours, and are sustained by the organism itself. What mechanisms make such cycles possible? It seems natural to look for an explanation in terms of chemical reactions. This model is adapted from J. Maynard Smith (1968, pp. 108–115). One of the simplest biochemical reactions that seems likely to offer an explanation is

1. A gene catalyzes messenger RNA (mRNA) production.
2. The mRNA leaves the nucleus and catalyzes the production of a protein.
3. A portion of the protein enters the nucleus and combines reversibly with the gene to form a product which does not produce mRNA.

Let M be the concentration of mRNA and P the concentration of protein. For simplicity we assume that there are many cells in the organism, that produce this protein, and so many copies of the relevant gene are present. Let G be the fraction of genes that are active, that is, not combined with the protein.

(*a*) The rate of the reaction

$$\text{Gene + protein} \longrightarrow \text{inactive}$$

is proportional to the product GP, and the rate of the reaction

$$\text{Inactive} \longrightarrow \text{gene + protein}$$

is proportional to $1 - G$. Show that the value of G at equilibrium is

$$G = \frac{1}{1 + aP}$$

for some $a > 0$. (You may wish to look at Problem 9.2.8.)

(*b*) Proteins and mRNA both decay. Defend the equations

$$\frac{dM}{dt} = \frac{b}{1 + aP} - cM,$$

$$\frac{dP}{dt} = eM - fP,$$

for some positive a, b, c, e, and f. Show that by suitably rescaling M, P, and t we can rewrite them as

(29)
$$m' = \frac{1}{1 + p} - \alpha m,$$

$$p' = m - \beta p,$$

for some positive α and β.

(c) It can be shown that (29) does not lead to sustained oscillations. In fact, no simple chemical reactions do, see Problem 9.2.8 and also J. S. Griffith (1968). One possible solution is to take into account the fact that it takes time for molecules to travel between the nuclei (where the genes are) and the sites where the protein is synthesized. Incorporate this into (29).

(d) Do the equations developed in (c) have sustained oscillations?

5. Walt Disney studios once filmed a simulated chain reaction which took place as follows. A large number of cocked mousetraps was placed on the floor of a bare room. Each trap was specially built so that when it was sprung it would throw two ping pong balls into the air. Flying ping pong balls that landed on unsprung traps would spring the traps and thereby set more balls flying. The reaction was started by tossing a single ping pong ball into the room. How should the simulation be designed so that the duration of the chain reaction will be reasonable— the audience must be able to see it, but it shouldn't last too long. The following treatment is adapted from G. F. Carrier (1966, pp. 2–6).

There are three obvious ways to influence the duration of the simulation: Change (1) the flight time of the balls, (2) the number of traps per square foot, or (3) the size of the room (keeping the number of traps per square foot the same by simultaneously changing the total number of traps). We consider each of these separately.

It can be observed that the flight times of the balls for a given brand of trap are nearly the same. We assume for simplicity that they're identical. After hitting a trap, very few balls are able to rebound enough to hit another trap with enough force to spring it. Thus a ball that hits a sprung trap or an unsprung trap becomes dead in most cases. We assume that this always happens. A ball that hits the bare floor may or may not rebound enough to be able to set off a trap; it depends on the flooring material. At any rate, there is a probability p that a random ball will land on a trap with enough force to spring it (if it is still cocked). The value of p depends only on how far apart the traps are and on the nature of the floor. (The latter is a fourth variable which we can adjust. You should convince yourself that this would have the same effect as changing the spacing of the traps.)

(a) Criticize the various assumptions we have made. What sorts of errors do they introduce into our predictions?

(b) Argue that the duration of the simulation is nearly proportional to the flight time of a ball. What advantages and disadvantages do you see in trying to adjust the duration by adjusting the flight time?

From now on, we use the flight time of a ball as the unit of time measurement.

(c) Let t be the length of time from the start of the simulation until b balls are in the air together, where b is much less than the total number of balls. Show that approximately $(2p)^t = b$, and so $t = \log b/\log 2p$. Consider two rooms in which the number of traps per square foot is the same, but one room is b times as large as the other. Show that the difference in the length of the simulations is about $\log b/\log 2p$. What advantages and disadvantages are there to adjusting the length of the simulation in this way? To what extent can you change the duration of the "middle" range—say the time to go from 5% sprung traps to 90% sprung traps? Discuss adjusting mousetrap density.

So far our discussion has dealt primarily with small t. Large t is harder. Intermediate t can be handled fairly easily. The rest of this problem is devoted to it.

(d) If there are N balls in flight at time n and U unsprung traps out of a total of M, show that the probability of having exactly $2B$ balls in flight at time $n + 1$, given that T of the traps are hit, is

$$P(B) = \left(\frac{U}{B}\right)\left(\frac{T}{M}\right)^B\left(1 - \frac{T}{M}\right)^{U-B},$$

where $\binom{U}{B}$ is the binomial coefficient "U choose B"— the number of ways to choose B objects from a set of U. Using the approximation that, if N is small compared to M, no trap is hit by more than one of the N balls, show that the probability that T traps will be hit is approximately

$$H(T) = \left(\frac{N}{T}\right)p^T(1 - p)^{N-T}.$$

(e) Describe a Monte Carlo simulation for the mousetrap demonstration. What inaccuracies have been introduced by our approximations?

(f) Both P and H can be approximated quite accurately by normal distributions for large values of U and N. The means and variances are

	Mean	Variance
For H	pN	$Np(1 - p)$
For P	UT/M	$UT(M - T)/M^2$

Let's consider the middle range of the experiment when U and N are both large. Show that, if N_n is the average number of balls in the air at time n, approximately,

$$(30) \qquad N_{n+1} = \frac{2pN_nU_n}{M}.$$

Since $U_{n+1} = U_n - N_{n+1}/2$, this can be solved recursively for N_n and U_n, but we can't see what's going on very well just by looking at (30).

(g) Write $f(n)$ for the fraction of unsprung traps at time n and show that (30) becomes

$$(31) \qquad f(n) - f(n+1) = 2p[f(n-1) - f(n)]f(n).$$

We approximate (31) by a differential equation in hopes of obtaining an easier problem. Replace $f(n+1)$ and $f(n-1)$ by their first degree Taylor polynomials about n. Show that this leads to $f'(n) = 2pf'(n)f(n)$, and so $f'(n) = 0$, a poor approximation. This means we need higher degree Taylor polynomials.

(h) Use quadratic Taylor polynomials to obtain the approximation

$$f'(n) + \frac{f''(n)}{2} = p[2f'(n) - f''(n)]f(n),$$

and so

$$(32) \qquad [pf(n) + \tfrac{1}{2}]f''(n) = [2pf(n) - 1]f'(n).$$

(i) Can you describe the solution to (32)? You cannot obtain an analytical solution, but (32) can be integrated once to obtain

$$f' = 2f - \frac{2}{p}\log(2pf + 1) + C.$$

(j) Using (31), (32), or some other device, find a way to answer the following questions. About how long does it take the simulation to go from $f(n) = 0.95$ to $f(n) = 0.1$? How large would you make p? Why?

THE HEUN METHOD

In case you have access to a computer but not to a library routine for solving differential equations, here is the Heun method for solving a system of first order equations of the form

$$y_i' = f_i(x, y_1, \ldots, y_n) = f_i(x, \mathbf{y}).$$

To take a single step of size h, set

$$\mathbf{y}^* = h\mathbf{f}(x, \mathbf{y}(x)) + \mathbf{y}(x),$$
$$\bar{\mathbf{y}} = h\mathbf{f}(x + h, \mathbf{y}^*) + \mathbf{y}^*,$$

and

$$\mathbf{y}(x + h) = \tfrac{1}{2}[\mathbf{y}(x) + \bar{\mathbf{y}}].$$

A check on the accuracy is provided by $\mathbf{y}^* - \mathbf{y}(x + h)$, which can be expected to be greater than the actual error. A better check is provided by using two values of h, since the error in integrating from $x = a$ to $x = b$ is roughly proportional to h^2. Thus, by using values of h differing by a factor of 2, we obtain two estimates for y, and their difference should be about three times the error obtained by using the estimate based on the smaller step size.

CHAPTER 9

LOCAL
STABILITY THEORY

If you wish a fuller discussion of the theoretical background than that presented here, consult a textbook. Some introductory differential equations textbooks contain a chapter or two on qualitative methods. F. Brauer and J. A. Nohel (1969) treat the general theory and discuss some specific problems.

9.1. AUTONOMOUS SYSTEMS

Suppose we are dealing with a system in which time is the independent variable. Absolute time may or may not appear. If absolute time appears, we are dealing with a *historical* system. If absolute time is irrelevant, the system is *autonomous*. Another way of looking at this is that the dependent variables are functions only of *differences* in time.

Suppose someone gives you money each day starting with \$1 today, \$2 tomorrow, \$3 the following day, and so on. Let the amount for day n be $M(n)$. Since you receive n dollars on the nth day, $M(n) = n$—a historical system. However, $M(n) = M(n - 1) + 1$—an autonomous system. Thus the distinction between historical and autonomous systems is sometimes artificial.

Here we are concerned only with the stability of autonomous systems and limit most of our discussion to systems with two first order equations involving two endogenous variables. This makes our discussion simpler, allows for two-dimensional diagrams, and still permits us to consider a variety of interesting models. Most of the mathematical ideas can be generalized to systems of higher order equations with several endogenous variables.

Suppose there is no time delay. Let the endogenous variables be $x = x(t)$ and $y = y(t)$. Since the equations are first order, we assume that they have been solved for x' and y' in terms of x and y, giving

$$(1) \qquad\qquad x' = f(x, y), \qquad y' = g(x, y),$$

where for the moment we do not say much about the functions f and g. Time can be completely eliminated from (1) by dividing one equation by the other to give

$$(2) \qquad\qquad \frac{dy}{dx} = \frac{g(x, y)}{f(x, y)}.$$

We can plot the solutions of the first order differential equation (2) in the xy plane. This is called the *phase plane*. Furthermore, an arrow can be attached to each curve indicating the direction of motion along the curve with time. This picture contains all the information in (1), except the rate of motion along the curves. (For n equations in n endogenous variables you can imagine the curves as lying in n-dimensional phase space.) The division of (1) to give (2) cannot be carried out if $f(x, y) = 0$ for some values of x and y. If $g(x, y) \neq 0$, the curve is vertical. If $g(x, y)$ is also zero (x, y) is called an *equilibrium point*. A solution that starts at an equilibrium point can never move, since $x', = y' = 0$ by (1). Such solutions are plotted simply as points.

By going from (1) to (2) we obtain a convenient way of representing solutions graphically. Also, (2) is usually more analytically tractable than (1). However, the loss of the time variable presents difficulties when we study stability questions.

There are two types of qualitative questions we can ask about the paths of solutions in the phase plane. If a solution starts near an equilibrium point, will it move toward the equilibrium point or away from it and in what manner? Questions of this type are dealt with in the subject area known as *stability in the small* or *local stability*. The second type of question does not assume that we start near an equilibrium point. It concerns what is known as *stability in the large* or *global stability*, a more difficult mathematical topic than local stability theory. I discuss this area briefly in Section 9.4. Global behavior is more varied than local behavior. For two first order equations the possibilities include divergence, convergence to an equilibrium point, periodicity, and convergence to a limit cycle. A *limit cycle* is a periodic solution such that a solution which starts nearby will approach it. (In the phase plane, a periodic solution appears as a simple closed curve.) In higher dimensions (i.e., three or more first order equations), global behavior is much more varied and much less understood.

9.2. DIFFERENTIAL EQUATIONS

Theoretical Background

The basic idea in local stability theory of differential equations is to approximate the system (1) by two linear first order differential equations near an equilibrium point. Suppose that (x_0, y_0) is an equilibrium point; that is,

$$(3) \qquad f(x_0, y_0) = g(x_0, y_0) = 0.$$

We want to approximate f and g near the point (x_0, y_0). Recall that for a function of a single variable, say $h(x)$, we can obtain a fairly good approximation near x_0 by using $h(x_0) + h'(x_0)(x - x_0)$ instead of $h(x)$. The same idea can be used with functions of two variables: We can approximate $f(x, y)$ near (x_0, y_0) by

$$f(x_0, y_0) + \frac{\partial f(x_0, y_0)}{\partial x}(x - x_0) + \frac{\partial f(x_0, y_0)}{\partial y}(y - y_0).$$

Here $\partial f(x_0, y_0)/\partial x$ denotes the partial derivative of f evaluated at the point (x_0, y_0), that is,

$$\lim_{t \to 0} \frac{f(x_0 + t, y_0) - f(x_0, y_0)}{t}.$$

To avoid cumbersome notation we denote this partial derivative by f_x. The meanings of f_y, g_x, and g_y should be obvious. Thus we have

$$(4) \qquad \begin{aligned} u' &\approx f_x u + f_y v, \\ v' &\approx g_x u + g_y v, \end{aligned}$$

where $u = x - x_0$, $v = y - y_0$, and f_x, f_y, g_x, and g_y denote the partial derivatives of f and g evaluated at (x_0, y_0). If we assume that the approximate equalities in (4) are exact, the equations can easily be solved. The solution of this homogeneous, linear system gives information about the local stability of the solutions of (1). Since our object is not to derive mathematical results, we merely state the following theorem which can be found in almost any differential equations textbook that discusses local stability theory.

THEOREM. If (x_0, y_0) is an equilibrium point for the system (1), define the real numbers b, c, and d by

$$b = \frac{f_x + g_y}{2},$$

$$c + di = b + \sqrt{b^2 - (f_x g_y - g_x f_y)},$$

where $i = \sqrt{-1}$.

If $c < 0$, the equilibrium point is *stable*; that is, solutions starting nearby move closer. If $c > 0$, the equilibrium point is *unstable*; that is, solutions starting nearby move further away. Furthermore, the distance from the equilibrium point behaves roughly like Ke^{ct}. If $c = 0$, additional tests will be needed to determine the nature of the equilibrium point. Necessary and sufficient conditions for $c < 0$ are $b < 0$ and $f_x g_y > g_x f_y$.

If $d \neq 0$, the solutions near the equilibrium point spiral about it in a roughly elliptical fashion with a period approximately equal to $2\pi/d$. The amplitude of the oscillation increases or decreases, depending on the sign of c. If $d = 0$, there is no oscillation.

Typical phase plane diagrams are illustrated in Figure 1 where it is assumed that (x_0, y_0) lies in the first quadrant.

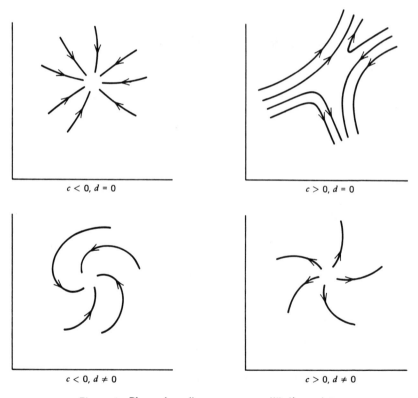

$c < 0, d = 0$

$c > 0, d = 0$

$c < 0, d \neq 0$

$c > 0, d \neq 0$

Figure 1 Phase plane diagrams near equilibrium points.

For those familiar with linear algebra, we note that c is the maximum of the real parts of the eigenvalues of the matrix $\|\partial f_i/\partial x_j\|$, where $x_1 = x$, $x_2 = y$, $f_1 = f$, and $f_2 = g$. Stated in this way, the stability result is valid for a system of n first order equations in n endogenous variables, but the nature of the oscillations is more complicated.

When we made the assumption that (4) is exact, we constructed a model of (1). Since the condition $c = 0$ is fragile, it is reasonable to suppose that we could not easily decide between stability and instability if $c = 0$. This is indeed the case. We do not study this situation here.

The condition $d = 0$ is equivalent to $b^2 - (f_x g_y - g_x f_y) \geq 0$ which can be put into the form

$$(5) \qquad d = 0 \qquad \text{if and only if} \qquad (f_x - g_y)^2 + 4f_y g_x \geq 0$$

by a little algebra. In particular, no oscillation occurs if $f_y g_x \geq 0$. Since $d = 0$ actually corresponds to an inequality, the case of oscillation versus nonoscillation is not fragile.

Frictional Damping of a Pendulum

Friction slows a pendulum down. It also changes its period. Will you need to allow for this change in designing a pendulum clock? If so, how?

We want to study the motion of a pendulum in an attempt to understand mathematically how frictional forces slow it down. These forces arise from the motion of the pendulum in the air, water, or whatever medium it is suspended in. In contrast to this, the motion of a frictionless pendulum is periodic. Using different methods, we studied the period of a frictionless pendulum in Section 2.2.

Consider a pendulum as shown in Figure 2. Since our primary interest is in damping due to friction, we make several simplifying assumptions whose removal is discussed briefly after the model is analyzed.

1. All the weight is concentrated as a point of mass m at the end of a piece of wire of length l. (If l is replaced by the distance from the pivot to the *center of mass*, the following results will remain valid.)
2. The wire does not stretch or wrap around its pivot, and so the length l is independent of the angle of the pendulum.
3. There is no wind, shaking, and so on, that can disturb the motion of the pendulum.

Let the angle of the pendulum be $\theta = \theta(t)$, where $\theta = 0$ is the rest position of the pendulum. The gravitational force acting on the pendulum is mg. It is partially balanced by tension in the wire. The resultant is the force

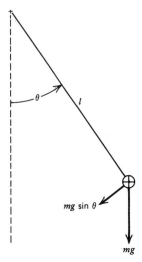

Figure 2   A pendulum.

$-mg \sin \theta$ acting on the pendulum along its direction of motion. The only other force affecting the motion of the pendulum is the frictional force we are examining. Empirical studies show that a function that depends only on the velocity gives a good approximation to such forces. Since the velocity of the pendulum depends on its angular velocity ω, we assume that the frictional force is of the form $-r(\omega)$, where r is a differentiable function. This is a retarding force, and so $r(\omega)$ has the same sign as ω. In particular, $r(0)$ is both nonnegative and nonpositive, and therefore is zero. We postpone further assumptions concerning the nature of r until they are needed. Newton's laws give

(6)
$$ml\theta'' = -mg \sin \theta - r(\omega).$$

We now show that our model predicts that friction causes the pendulum to slow down. Since $\omega = \theta'$, we can rewrite (6) as

(7a)
$$\theta' = \omega$$

(7b)
$$\omega' = -\frac{g}{l} \sin \theta - \frac{r(\omega)}{lm}.$$

These equations are in the form (1), with $x = \theta$ and $y = \omega$. We set (7) equal to zero to find the equilibrium points. From (7a) we have $\omega = 0$. Since $r(0) = 0$, we deduce from (7b) that θ is a multiple of π. Because of the periodicity of the sine function, we need only consider $\theta = 0$ and $\theta = \pi$. The latter case corresponds to the pendulum being straight up. It is left as an exercise to show that this equilibrium point is unstable.

We have $(\omega_0, \theta_0) = (0, 0)$. The partial derivatives are

(8)
$$(\theta')_\theta = 0, \qquad (\theta)_\omega = 1,$$
$$(\omega')_\theta = \frac{-g}{l}, \qquad (\omega)_\omega = \frac{-r'(0)}{lm}.$$

Since $r(\omega)$ is an increasing function near $\omega = 0$, it is reasonable to assume that $r'(0) > 0$ (This really is an assumption; consider $r(\omega) = \omega^3$.) In the theorem we have

$$b = \frac{-r'(0)}{2lm},$$

$$c + di = b + \sqrt{b^2 - \frac{g}{l}}.$$

Since $\sqrt{b^2 - g/l} < |b|$, $c < 0$. It follows that the motion of a pendulum is locally stable; that is, it dies out. We see that d is nonzero if $b^2 < g/l$, which can be rewritten as $r'(0) < 2m\sqrt{gl}$. Hence the pendulum oscillates if $r'(0) < 2m\sqrt{gl}$ and does not oscillate if $r'(0) > 2m\sqrt{gl}$. In the latter case the frictional forces are very large, and it is as if the pendulum were moving in molasses.

How does the change in the period of the pendulum compare with the damping? From the theorem, the pendulum will oscillate at about half its initial amplitude after a time t, where $e^{ct} = \frac{1}{2}$. Hence $t = \log_e (2)/|b|$. This requires about $t/(2\pi/d)$ oscillations of the pendulum. Hence the pendulum loses half its amplitude after about

(9)
$$\frac{\log (2) d}{2\pi |b|} = \frac{0.11032\, d}{|b|} = 0.11032 \sqrt{\frac{g}{lb^2} - 1}$$

oscillations. Call this number n. Squaring and rearranging, we obtain $g/lb^2 \approx 82n^2 + 1 \approx 82n^2$. The ratio of the period of the pendulum to the period of a frictionless pendulum is

(10)
$$\left(1 - \frac{lb^2}{g}\right)^{-1/2} \approx 1 + \frac{lb^2}{2g} \approx 1 + \frac{1}{164n^2}.$$

Thus the period increases by about $0.6/n^2$ percent, a very small change. This prediction can be tested experimentally.

Since a pendulum takes quite a long time to slow down in air, n is large in this case. It follows that the effect of friction on the period is quite small. If this were not so, the period of a pendulum would depend on barometric pressure and pendulum clocks would not keep accurate time.

It is possible to replace (6) by the more general equation $ml\theta'' = f(\theta, \omega)$, where we assume only that f_ω and f_θ are both negative near zero as they are in (8). Since $f_\theta = -mg$ is a good approximation, our conclusions are unchanged.

Species Interaction and Population Size

Interaction between species lies at the heart of ecology. Some claim that these interactions cause the nearly cyclic fluctuations observed in some populations. Others claim that other factors are responsible. What can a simple mathematical model contribute to the debate?

Since our theorem allows only two endogenous variables, we assume only two species are interacting. Let x be the number of organisms in the first species and y the number of the second. There are three basic types of interaction between species:

1. The first species preys on the second (either direct predation or as a parasite).
2. Both species compete for more or less the same limited resources (e.g., plants competing for sunlight).
3. The two species live in a symbiotic relationship with each other (e.g., nitrogen fixing bacteria on the roots of peas and beans).

Predation is discussed below. Competition and symbiosis are treated sketchily, and the details left as exercises.

The assumption of autonomy implies that the environment of the species is constant except for factors whose change depends only on the number of organisms of the two types. Because of the form of (1), no time lag can be used. Since species require time to reproduce, the absence of a time lag may be a serious deficiency. Furthermore, the past history of the population determines the age mix and general physical condition of the present population. It is an open question how serious a restriction avoiding the past is. If it is serious, difference equations or mixed differential difference equations will be needed. See R. M. May (1973) for a relevant discussion.

Let x be the number of predators and y the number of prey. It is intuitively clearer to think in terms of the net growth rates of the two species:

(11) $$\text{Predator}: \frac{x'}{x} = r(x, y), \qquad \text{Prey}: \frac{y'}{y} = s(x, y).$$

The equilibrium condition is then $r = s = 0$. A historically important special case is the Volterra–Lotka equations:

(12) $$r(x, y) = a + by, \qquad s(x, y) = c + dx.$$

J. G. Kemeny and J. L. Snell (1962, Ch. 3) discuss this special model.

We deal with a more general model in which r and s are only vaguely specified. To be able to say something about the stability of the system, we must make some assumptions about r and s.

If the population size of species 1 does not affect the population growth of species 2, then $r_y \equiv 0$. Similarly, if species 1 does not affect its own population growth through crowding, resource exhaustion, and so on, we will have $r_x \equiv 0$. Because the absence of an effect leads to zero for the partial derivative, an indication of how the species affect one another gives us information about the signs of the various partial derivatives. We may be able to make educated guesses about their relative magnitudes as well. Actual data collection is very difficult at best.

What effect does a change in predator population have on the net growth rate of the prey? Since predators consume prey, $s_x < 0$. If the predator population increases, there will be less food per predator, and so $r_x < 0$. Another way to decrease the number of prey per predator is to reduce the the prey population, so we expect $r_y > 0$. The sign of s_y is harder to determine. as the prey population increases in the absence of predation, the net growth rate should decrease, because the species is now moving into less favorable parts of the environment. However, if the prey increase while the predators do not, there will be less predator pressure per individual of prey population, which would lead to an increasing net growth rate. These two effects tend to cancel out. Note that, if the predators are prey at a higher level in the food chain, the argument just given for s_y also applies to r_x. We have reached the following conclusions:

(13) $$r_x < 0 \text{ or } \approx 0, \qquad r_y > 0,$$
$$s_x < 0, \qquad s_y \approx 0.$$

We interpret $s_y \approx 0$ to mean that we can neglect s_y compared to the other partial derivatives.

Suppose there is an equilibrium point (x_0, y_0) at which neither species has vanished. From (11) we have $r = s = 0$. The question of the existence of solutions to this equation is discussed later. It is easily seen that $f_x = x r_x$, $f_y = x r_y$, and so on. From the theorem and (5) we have

(14) $$b = \frac{x r_x + y s_y}{2}$$
$$c + di = b + \sqrt{b^2 - xy(r_x s_y - r_y s_x)}.$$

From (13) we see that b is negative and $r_x s_y > r_y s_x$. Thus we have local stability. If the environment is rather homogeneous, the self-limiting effect of species 2 will not come into play until the environment is nearly saturated. In this case we have $s_y > 0$, and instability is possible. Hence heterogeneity of the environment increases stability. See R. M. May (1972) and M. L. Rosenzweig (1971).

Do we have oscillation? From (5) we see that the answer is yes if and only if

(15) $$(xr_x - ys_y)^2 + 4xyr_y s_x < 0.$$

This holds if $r_x \approx 0$. Roughly speaking, (15) says that the interspecies effects on net birth rates are greater than the intraspecies effects. This certainly appears to be true in some situations.

Unless we are willing to make a statement stronger than (13) we can't really say much more; however, this is quite a lot considering the vagueness of (13).

We now turn our attention to the existence of equilibrium points (x_0, y_0). The arguments used to derive (13) did not use the assumption that we were at equilibrium, so we drop the assumption that the partial derivatives in (13) are evaluated at equilibrium. Another point is that, in deriving $s_y \approx 0$, we used the fact that the predator was severely limiting the prey; let's relax this by allowing $s_y < 0$ at low predator densities. The following discussion is entirely nongraphical. In simple cases such as this one, a graphical discussion may be preferable. You are asked to provide this in Problem 2.

To begin with, there are the trivial equilibrium points associated with $x = 0$ (no predators). We put them aside and look for equilibrium points with predators. Hence $x > 0$ and $y > 0$.

Suppose that x predators can survive if there are enough prey and if the number of predators does not exceed some critical value x_m. Since $r_y > 0$, we can solve $r(x, y) = 0$ for a unique $y(x)$. By implicit differentiation $dy/dx = -r_x/r_y$, which is positive by (13). If predators cannot live in the absence of prey, we have $y(x) > 0$ for $0 < x \le x_m$. Substituting $y(x)$ in $s(x, y)$ we find

$$\frac{ds}{dx} = s_x + s_y \frac{dy}{dx},$$

which is negative because of (13).

Now we can show that a unique equilibrium exists under certain conditions. We have just shown that $s(x, y(x))$ is a strictly decreasing function of x. Therefore an equilibrium point will exist if and only if $s(x, y(x))$ is positive for small x and negative for large x. If such a point exists, it will be unique because $s(x, y(x))$ is *strictly* decreasing. To say that $s(x, y(x)) > 0$ means that we must *remove* prey to keep the prey population from increasing beyond $y(x)$. This is likely to be true for small x and false for large x, which is just what we want. If the predator population is kept rather low (small x_m) by exogenous forces such as hunting by humans, the prey may be able to provide food for the predator and still increase. In that case we cannot reach the portion of the curve $y(x)$ where $s(x, y(x)) \leq 0$. Discuss this situation.

Let's relate our conclusions to the real world.

The main result of our study is a model which proposes a mechanism for maintaining stability in a world that is changing. If the environment varies a lot in a time period comparable to that in which x and y move significantly toward equilibrium, our results will be useless. However, infrequent changes can be viewed as occasional displacements from equilibrium; for example, a change in the environment actually shifts the location of the equilibrium point by changing the functions r and s; an infrequent epidemic changes the value of x or y but leaves the equilibrium point unchanged. If these displacements are not too large, our use of local stability theory shows that the system will tend to return to equilibrium. If the system possesses global stability, even large displacements will be damped out. See R. M. May (1973).

Most natural systems involve many predator and prey species. If we introduce one variable for each species, much more than the vague conditions in (13) will be needed to study stability. What can be done about this? If the prey species are sufficiently alike, we can lump them together as if they were one species. Likewise for the predators. In this way it may be possible to apply our conclusions to a system involving more than two species. Since the model would only make predictions about the size of the lumped species population, the individual populations may fluctuate wildly.

How can we gather data to test the model? Except in the physical sciences or in carefully controlled experiments, it is usually difficult to estimate first derivatives and nearly impossible to estimate higher derivatives. Therefore we should not try to verify (11) and (13) directly. To check the model we need some predictions that can be feasibly tested. The model predicts that the population sizes will exhibit damped oscillations with nearly constant periods if they are disturbed from equilibrium. If (4) is treated as an equality and solved, it can be shown that the relative maxima of u and v differ by a constant phase. As a result, we predict that the predator and prey cycles will be out of phase with one another by about the same amount from cycle to cycle. Of course, random disturbances cause variations, so neither of these predictions is perfect. We now have two predictions which it may be feasible to check: nearly constant period and nearly constant phase shift.

In the last section of their article N. S. Goel et al. (1971) briefly discuss some experiments that have been done to check the model. The predictions are usually correct.

It is difficult to test the model on natural populations. The most well-known candidate is the system consisting of the Canadian lynx and the snow-shoe hare. Since it undergoes wild fluctuations, a global result is needed. However, the hare population fluctuates in the absence of lynx predators, so a simple lynx–hare model is wrong. Furthermore, the relative phases of the lynx and hare fluctuations seem wrong (Gilpin, 1973). A more promising model may be some sort of three-way system involving hares, vegetation, and (exogenously) the weather. For further discussion of this problem see L. B. Keith (1963).

We consider competition and symbiosis briefly. An important factor in the competition situation is the existence of an equilibrium point. This is discussed in Problem 3.3.3. For competition all the partial derivatives are negative. There will be stability if and only if $r_x s_y > s_x r_y$. Roughly speaking, this says that each species inhibits its own expansion more than the competing species does.

We now turn our attention to symbiosis. Assume that, if one species somehow increases, it will help the other to increase too. This means that r_y and s_x are positive. It follows from (5) that $d = 0$, and so there is no oscillation. Suppose we increase species 1 by a small percentage. Ignoring self-limitation, this is essentially the same as decreasing species 2 by the same percentage. Thus we expect $xr_x \approx -yr_y$, unless species 1 tends to limit itself. Self-limitation makes r_x an even larger negative number, and so $|xr_x| > yr_y$ in this case. If a similar result holds for the second species, the equilibrium will be stable, because stability is equivalent to $r_x s_y > s_y r_x$.

Keynesian Economics

J. M. Keynes's revolutionary work, *The General Theory of Employment, Interest and Money*, has had a profound effect on economic theory and practice, the latter beginning with Roosevelt's New Deal politics during the U.S. Depression of the 1930s. Here we study a crude bare bones model adapted from G. Gandolfo (1971).

Let's begin with a list of variables that relate to the national economy:

C, desired level of consumption.
I, desired level of capital investment.
D, total demand for goods.
Y, national income.
L, desired amount of money to be held as cash on hand.
M, amount of money available.
R, cost of money (interest rate).

I'm sure you can add to the list, but we have enough for the time being. You may wish to come back later and add more.

Before discussing these variables, a word about measurement is worthwhile. Some of these quantities may be hard to measure, partly because they are imprecisely defined and partly because it is not clear what units we should use. Although lack of precision is a serious problem, we ignore it here because we can construct a model without it and because attempting to eliminate it would involve us in deep economic considerations. We want to measure our variables in *real terms*, whatever this slippery phrase means. Economists use *constant dollars*, that is, dollars deflated to some standard year such as 1950. We avoid problems by assuming that our variables are somehow measured in constant dollars (except for R which is a ratio of constant dollars). Note that R is negative if the rate charged by moneylenders is less than the rate of inflation. All this lack of precision is really a serious problem. If it is not resolved, two people may mean different things by the same terms and so the discussion of models will become hopelessly muddled. This may be part of the problem at the present time. People are arguing over whether or not the current (1974) combination of high unemployment and inflation (called *stagflation*) shows that Keynesian models cannot be used.

Back to our model.

Capital investment over a period of time increases the efficiency of labor. To avoid this thorny problem, we deal with a short term model, that is, one in which the change in total capital investment is not significant. Technological development creates a similar problem which we also avoid by using a short term model.

Having said what we won't try to do, let's see what we can do. Our list of variables is too long to handle easily, so we need to know which are exogenous (independent), which are endogenous (dependent), and which we can ignore. Unfortunately, to do this sort of thing directly can be very difficult; it is often easier to sneak up on it through discussion.

At equilibrium we will have $D = Y$ and $L = M$, that is, what we want is what we have. Since we are concerned with disequilibrium, the quantities $D - Y$ and $L - M$ are important. Since excessive demand for money drives up the interest rate and excessive demand for goods causes production to increase, we assume that

$$(16) \qquad \begin{aligned} R' &= r(L - M), & r'(0) &> 0, & r(0) &= 0, \\ Y' &= y(D - Y), & y'(0) &> 0, & y(0) &= 0. \end{aligned}$$

This suggests that it would be nice if we could take R and Y as the basic variables, influencing their own growth through (16). Can we relate L, M, D, and Y to R and Y? Of course, Y presents no problem: $Y = Y$.

Since M is determined by the government, it is an exogenous variable. We assume that it is constant for the purposes of studying stability; however, it is interesting to ask how changes in M influence the equilibrium value of Y; that is, what is the sign of $Y_M = \partial Y/\partial M$ at equilibrium? Back to this later.

What about L? It seems reasonable to assume that $L_R < 0$ and $L_Y > 0$, since people want to hold less cash as interest rates rise and the country needs more cash for transactions as national income rises. This is far from an explicit functional relationship, but we'll see how far we can go with it. This approach worked fairly well in the previous model.

To study the partial derivatives of D, it is convenient to break it into two parts: $D = C + I$. The value of C_R should be zero or negative, since higher interest rates should, if anything, be an inducement to save. Can you defend the assumption $C_Y > 0$? What about $I_R < 0$ and $I_Y \geq 0$? It follows that $D_Y > 0$ and $D_R < 0$.

In summary,

$$\text{(17)} \qquad \begin{array}{llll} L_R < 0, & D_R < 0, & I_R < 0, & C_R < 0, \\ L_Y > 0, & D_Y > 0, & I_Y \geq 0, & C_Y > 0. \end{array}$$

We now compute the partial derivatives of r and y at equilibrium:

$$\text{(18)} \qquad \begin{array}{ll} r_R = r'(0)L_R < 0, & r_Y = r'(0)L_Y > 0, \\ y_R = y'(0)D_R < 0, & y_Y = y'(0)(D_Y - 1). \end{array}$$

The sign of y_Y cannot be determined from (17). The stability conditions in the theorem are

$$r_R + y_Y < 0 \qquad \text{and} \qquad r_R y_Y > r_Y y_R.$$

A sufficient, but not necessary, condition for this to hold is $y_Y \leq 0$. In words,

> If the sensitivity of total demand to changes in the national income is less than unity, our Keynesian model is locally stable.

(In economics "sensitivity" is a term for a partial derivative; the sensitivity of A to B is the amount A changes when B changes one unit.) When does the proposition stated above apply? We must have $C_Y < 1$ to ensure $D_Y < 1$. What does $C_Y > 1$ mean? It says that, as income increases, desired consumption increases even faster; an unlikely possibility except in underdeveloped countries where it can cause severe problems. (See Problem 3.3.5.) We can't use this argument on $C + I$, because consumers and investors do not consult each other. However, I may not be at all sensitive to Y—it's Y' that we can expect I to depend upon, since *changes* in Y stimulate additional investment (if Y increases) or liquidation (if Y decreases). As a first approximation, $I_Y = 0$, and so the hypothesis of the proposition is satisfied.

Now suppose that we have stability. How will the equilibrium move in response to government adjustment of the money supply? At equilibrium, $L = M$ and $D - Y = 0$. Using the chain rule to differentiate these with respect to M we obtain

$$L_R R_M + L_Y Y_M = 1 \quad \text{and} \quad D_R R_M + (D_Y - 1)Y_M = 0.$$

Thus

(19) $$R_M = \frac{D_Y - 1}{\Delta} \quad \text{and} \quad Y_M = -\frac{D_R}{\Delta},$$

where

$$\Delta = (D_Y - 1)L_R - D_R L_Y.$$

Comparing (18) and (19), we see that the stability condition $r_R y_Y > r_Y y_R$ is equivalent to $\Delta > 0$. Since we are assuming stability, $\Delta > 0$. By (17) and (19), $Y_M > 0$. If the hypothesis in the proposition is true, $R_M < 0$. Government often tries to influence national income by adjusting M, for example, by making more money available when unemployment is high. (Making money available is not simply a matter of running the printing presses—this only leads to inflation with little change in the money supply as measured in real dollars. In the United States the Federal Reserve Board changes the percentage of cash reserves that member banks must hold.) What effect does this have? Since $Y_M > 0$, this should increase national income. Because of our assumption that we are dealing with the short term, national income can increase only by an increase in labor. Hence unemployment should decrease. The size of the change depends on the change in M and the size of Y_M. If D_R is small, we see from (19) that changing M may not be a very effective way to fight large scale unemployment.

Governments try other methods of influencing the economy, which may be more effective than controlling M. Can you change the model to allow for government control of R? What about direct attempts to influence D through deficit spending? Can you extend the model to allow for effects of taxation? Taxation can influence C and I by redistributing Y and by providing tax incentives for investment.

More Complicated Situations

B. Noble (1971, Ch. 6) presents two engineering applications: one in hydrodynamics and the other in chemical engineering. I have limited the material in this section to two first order equations. T. V. Kármán and M. A. Biot (1940, pp. 249–255) use two second order equations to discuss the stability of an airplane. L. S. Pontryagin (1962, pp. 213–220) uses three equations to discuss the stability of a stream engine governor. N. Rashevsky (1964, Part IV) uses various numbers of equations to discuss endocrine systems. He assumes the equations are linear. Instead, one can apply local stability theory to equations of a fairly general form.

PROBLEMS

Problems 1 and 2 deal with the predator-prey model, but do not use local stability theory.

1. Study the existence of equilibria in the predator-prey model graphically by plotting the two curves $x' = 0$ and $y' = 0$. Limit yourself to $x > 0$ and $y > 0$, and use (13) to help determine slopes.

2. The gypsy moth caterpillar causes considerable damage to trees. Consider a predator-prey model in which the prey is the gypsy moth and the predator is one of several parasitic wasps that attack gypsy moth caterpillars. Since the wasp larvae feed on gypsy moth caterpillars, killing the caterpillar also kills the wasp larvae. A spray program is instituted for gypsy moth caterpillars, using a general purpose insecticide. Suppose that the result is an increase in the death rate of gypsy moths and wasps by an amount ρ independent of the number present.

   (a) Is this a reasonable approximation? Why?

   (b) Using the results of the previous problem, predict the effect of the moth control program on the equilibrium size of the wasp population. Show that more data are needed to predict the effect on the moth population.

   (c) Let x_0 and y_0 be the solutions of the equations $r(x, y) - \rho = 0$ and $s(x, y) - \rho = 0$. Compute $dx_0/d\rho$ and $dy_0/d\rho$ and show that they have the same signs as $s_y - r_y$ and $r_x - s_x$, respectively, without using (b).

   (d) Use (c) and (13) to verify the graphical conclusions derived in (b).

   (e) Suppose that the wasps have little effect on the size of the gypsy moth population. This is probably the case when the gypsy moth population suddenly explodes. (Why?) Show that in this case *spraying will cause the gypsy moth population to decrease.*

   (f) Suppose that the gypsy moth is limited by the parasite rather than by intraspecies competition. This is probably the case when the gypsy moth population is fairly stable. (Why?) Show that in this case *spraying will cause the gypsy moth population to increase.*

This model has rather interesting implications for insecticide usage policies. The following experience agrees with the prediction in (f). In 1868 the cottony cushion scale insect was introduced into the United States from Australia and began to attack citrus groves. The ladybird beetle was introduced afterward as a predator to control the pest. When the citrus industry later tried to use DDT to reduce the scale population further, the number of pests actually increased (N. S. Goel et al., 1971).

Because the gypsy moth population undergoes wild swings, I have doubts about the accuracy of the above predictions. However, the model does indicate some problems that must be considered in planning a control program.

(g) The following data refer to percentages of true fish in catches brought into the port of Fiume, Italy. The remainder of the catch (sharks, rays, etc.) were primarily predators which feed on true fish. Can you explain the data? Note that during World War I, which ended in 1917, the amount of fishing was below peacetime levels. The data come from M. Braun (1975) who obtained them from the work of U. d'Ancona.

1914	1915	1916	1917	1918	1919	1920	1921	1922	1923
88%	79%	78%	79%	64%	73%	84%	84%	85%	89%

3. Develop the symbiosis model for species interaction.

4. The Keynesian model involves a variety of functions. Can you describe some of the graphs associated with them? In particular, what does the Y-R phase plane look like? You need to graph $D = Y$ and $L = M$ in the Y-R plane.

5. (a) Suppose we replace (16) in the Keynesian economics model by the more general equations $Y' = y(D, Y)$ and $R' = r(L, M)$. What can you say about the form of y and r? Do our conclusions remain valid?

   (b) In the Keynesian model we could include sensitivity of investors to changes in Y and R, that is, $I(Y, R, Y', R')$. Can you say anything useful about such a model?

6. In this problem we consider the armaments of two antagonistic countries or blocs. Suppose that (1) provides an adequate description of the amount of armaments x and y of the two antagonists. Allow for maintenance costs and the pressure for higher or lower armament levels provided by the opponent's arms level. Discuss the behavior of the

model. Can you interpret negative values for x and y? You may have to introduce a new definition for x and y in place of armament level. Perhaps something like the level of aggressiveness would work. How much faith do you have in the predictions you have made?

The linear form of this model was introduced by Richardson. He showed that it provided a good fit to European data from 1909 to the outbreak of World War I. See L. F. Richardson (1960) or T. L. Saaty (1968, pp. 46–48) for further discussion.

7. Apply the methods of this section to the group dynamics model of Section 3.3.
8. If various chemicals are reacting in a closed system (i.e., nothing can be removed or added), reactions often stop before any of the chemicals are completely exhausted. Can this stable equilibrium be explained simply in terms of the basic model for chemical reactions? [By "basic model" I mean the mass action model developed below in (b).] Let the various chemicals present be denoted by X_i. Suppose that m_1 molecules of X_1 plus m_2 molecules of X_2, and so on, can react to produce n_1 molecules of X_1 plus n_2 molecules of X_2, and so on. We assume that the reaction is reversible. This is written in the form

$$\sum_i m_i X_i \quad \longleftrightarrow \quad \sum_i n_i X_i.$$

In (a) through (c), we assume that this is the only reaction that is occurring.

(a) Let $C_i(t)$ be the concentration of chemical X_i at time t. Show that

$$C_i(t) = C_i(0) + (n_i - m_i)x(t),$$

where $x(t)$ is some function independent of i. How can $x(t)$ be interpreted?

(b) Suppose that a reaction can occur only if m_i molecules of X_i all collide with one another simultaneously. Conclude that the forward reaction ($\rightarrow$) proceeds at the rate

$$k_f \prod_i C_i(t)^{m_i},$$

where k_f is a constant called the *rate constant* for the forward reaction. Let k_b be the rate constant for the backward reaction. Show that the equation for the reaction is

$$x'(t) = k_f \prod_i C_i(t)^{m_i} - k_b \prod_i C_i(t)^{n_i}.$$

(c)   Conclude that the chemicals are in equilibrium if and only if

(20)
$$k_f \prod_i C_i(t)^{m_i} = k_b \prod_i C_i(t)^{n_i}.$$

Often several reactions occur at once. We want to show that the equilibrium points determined by (20) are locally stable. Only two simultaneous reactions are considered because of the limitations imposed in this section; however, the approach and results hold for any number of reactions.

(d)   Repeat the analysis in (a) through (c) assuming that the two reactions are

$$\sum_i m_i X_i \quad \longleftrightarrow \quad \sum_i n_i X_i,$$

$$\sum_i p_i X_i \quad \longleftrightarrow \quad \sum_i q_i X_i,$$

with rate constants k_f, k_b, r_f, and r_b; introduce $x(t)$ and $y(t)$ associated with these reactions so that

$$C_i(t) = C_i(0) + (n_i - m_i)x(t) + (q_i - p_i)y(t),$$

$$x'(t) = k_f \prod_i C_i(t)^{m_i} - k_b \prod_i C_i(t)^{n_i},$$

$$y'(t) = r_f \prod_i C_i(t)^{p_i} - r_b \prod_i C_i(t)^{q_i}.$$

(e)   Denote the four products, *including the rate constants*, appearing in the above formulas by $\pi(m)$, $\pi(n)$, $\pi(p)$, and $\pi(q)$, respectively. Show that equilibrium occurs if and only if $\pi(m) = \pi(n)$ and $\pi(p) = \pi(q)$.

(f)   Write $x' = f(x, y)$ and $y' = g(x, y)$. Show that at equilibrium

$$f_x = -\pi(m) \sum_i \frac{(m_i - n_i)^2}{C_i(t)},$$

$$f_y = -\pi(m) \sum_i \frac{(m_i - n_i)(p_i - q_i)}{C_i(t)},$$

$$g_x = -\pi(p) \sum_i \frac{(m_i - n_i)(p_i - q_i)}{C_i(t)},$$

$$g_y = -\pi(p) \sum_i \frac{(p_i - q_i)^2}{C_i(t)}.$$

(g)   Show that $f_x + g_y$ is negative. Use the Cauchy–Schwartz inequality

$$\left(\sum W_i^2\right)\left(\sum Z_i^2\right) \geq \left(\sum W_i Z_i\right)^2$$

to show that $f_x g_y \geq f_y g_x \geq 0$.

(*h*) Discuss the behavior of the reactions near equilibrium.

D. Shear (1967) establishes global stability under fairly general conditions.

9. Apply the methods of this section to the graduate student model in Problem 3.3.2.

10. In this problem you will study models for gonorrhea epidemics. For more material on epidemics see N. T. J. Bailey (1976). Gonorrhea is spread by sexual intercourse, takes 3 to 7 days to incubate, and can be cured by the use of antibiotics. Furthermore, there is no evidence that a person ever develops immunity.

(*a*) Let x be the fraction of men who are infected and let f be the fraction of men who are promiscuous. Let X and F be the corresponding quantities for women. Discuss the model

$$x' = -ax + b(f - x)X,$$
$$X' = -AX + B(F - X)x,$$

where a, b, A, and B are constants. Interpret the constants.

(*b*) What are the equilibrium points of this model? Which ones are stable? Provide phase plane sketches. You should find that the number $(a/bf)(A/BF)$ is critical. When will there be a continual epidemic?

(*c*) Interpret and discuss the effects of changes in the frequency of promiscuous intercourse, the fraction of the population (of either sex) that is promiscuous, and the speed of curing infections.

(*d*) What advice would you give to public health officials who wished to stem a gonorrhea epidemic in an affluent country like the United States? In a place like Hong Kong?

(*e*) Develop a model like the above for a population of male homosexuals. Such a model may be applicable to diseases not linked to sex, for examples, measles and typhoid. See Problem 8.1.3.

(*f*) Develop a less specific model; for example,

$$x' = g(x, X) \quad \text{and} \quad X' = G(x, X),$$

with minimal assumptions about g and G.

(*g*) Can you apply any of the above ideas to diseases that require two hosts? An example is malaria which is transmitted by mosquitoes.

9.3. DIFFERENTIAL DIFFERENCE EQUATIONS

We now briefly consider equations involving both derivatives and time lags. As in the previous section, we expand the equation(s) around an equilibrium point to obtain a homogeneous linear approximation. These approximations can be studied with Laplace transforms. We describe an alternative approach which is equivalent to this but does not require a knowledge of Laplace transforms.

For simplicity assume that there is only one equation in one endogenous variable. Write Taylor's theorem in the form

$$(21) \qquad f(t + \tau) = \sum_{n=0}^{\infty} \frac{(\tau D)^n}{n!} \; f(t) = e^{\tau D}f(t),$$

where D stands for d/dt. We could use this, for example, to rewrite $f'(t) = bf(t - 1) - mf(t)$ as $(D - be^{-D} + m)f(t) = 0$. In this way any homogeneous linear differential difference equation can be replaced by an infinite order differential equation $L(D)f(t) = 0$, where the function L is a polynomial in D and $e^{\tau D}$ for various values of τ. If the equation was of finite order, the general solution would be a linear combination of solutions of the form $t^n e^{rt}$, where r is a root of $L(r) = 0$ of multiplicity greater than n. The stability of the equation could then be determined by looking at the roots of $L(r) = 0$. (Section 9.2 dealt with quadratic L, because eliminating v and v' from (4) leads to one second order differential equation.) This method also works for the infinite order equation.

Since $L(r) = 0$ is a transcendental equation, studying its roots is often very difficult. A computer may be essential. There are usually an infinite number of roots, so it is fairly likely that at least one will have a positive real part. Hence local instability is common.

I can't resist the side remark that (21) can also be used to derive numerical integration and differentiation formulas. For examples see L. P. Ford (1955, Ch. 8).

The Dynamics of Car Following

Traffic flow has become the subject of mathematical modeling in recent years. Three authors who discuss it are W. D. Ashton (1966), F. A. Haight (1963), and L. J. Pignataro (1973). Sometimes cars are considered individually, and systems of equations or probabilistic models are developed. At other times traffic is treated as a fluid, and hydrodynamic techniques are used. Among the topics considered in traffic flow are the motion of traffic on the open road, bottlenecks, and effects of intersections.

How do drivers in a line of cars behave? There is a limit to how fast a driver can react, but too much delay in reacting causes collisions. Are the delays in drivers' reactions near the danger level? The model is adapted from R. Herman et al. (1959) and R. E. Chandler et al. (1958), which use the Laplace transform method. That approach is adapted for use as a student project by E. A. Bender and L. P. Neuwirth (1973). Related material appears in the first part of J. Almond (1965).

The driver of a car cannot directly control the speed of the vehicle. Instead, he or she controls its acceleration. Thus we expect to derive a formula for the acceleration as a function of the driver's sensitivity and the stimulus of the environment. Historically the model has been taken to be of the form

(22)   Acceleration = Sensitivity × Stimulus.

Since we have not defined what we mean by either "sensitivity" or "stimulus," the above formula has no content. Rather than attempt to give meaning to the terms "sensitivity" and "stimulus," we consider directly the physical factors that enter into the driver's reaction.

The driver's reaction (acceleration) depends on what he or she senses in the environment. The things that can be perceived most easily are the car's speed, its speed relative to other cars in the line, and the space between the car and adjacent cars. As an approximation, we suppose that the only relevant car is the one directly ahead of the driver. If x_n denotes the position of the nth car, we can write

(23)   Acceleration $= f(x_n', x_{n-1}' - x_n', x_{n-1} - x_n).$

In order to proceed it is necessary to say something about the nature of f. Experimentation seems to indicate that the most important factor is the relative velocity. To begin with we construct the simplest possible model using this: Acceleration is directly proportional to the relative velocity.

There is a delay, called the *reaction time*, between a change in the environment and the driver's response. It has been observed to be of the same order of magnitude as the time it takes the vehicle to cover the distance between it and the car ahead. Hence we expect the reaction time to be an important variable. To check this we compare the resulting model with one lacking a reaction time.

Let τ_n be the reaction time of the nth driver. The above discussion leads to the basic equation

(24) $x_n''(t + \tau_n) = \lambda_n[x_{n-1}'(t) - x_n'(t)],$

where λ_n is a constant measuring the strength of the nth driver's response. Chandler et al. (1958) conducted an experiment on the General Motors test track in which one driver followed another at what was considered to be a minimum safe distance. Equation 24 gave a good fit for most of the drivers when statistical methods were used to estimate λ and τ. These parameter estimates are given in Table 1.

Table 1   Driver Reaction Parameters.

Driver Number	τ (seconds)	λ (sec^{-1})	$\tau\lambda$
1	1.4	0.74	1.04
2	1.0	0.44	0.44
3	1.5	0.34	0.51
4	1.5	0.32	0.48
5	1.7	0.38	0.65
6	1.1	0.17	0.19
7	2.2	0.32	0.70
8	2.0	0.23	0.46

Source: Chandler et al. (1958).

We can rewrite (24) in operator notation as

$$\left(1 + \frac{De^{\tau_n D}}{\lambda_n}\right)v_n = v_{n-1},$$

where $v = x'$ is the velocity of the car. Using the subscript 0 to denote the lead car, we have

$$(25) \qquad \left[\prod_{k=1}^{n}\left(1 + \frac{De^{\tau_k D}}{\lambda_k}\right)\right]v_n = v_0.$$

We get our stability information from this equation.

To apply local stability theory we assume that $v_0(t)$ is given and that a stable particular solution $v_{np}(t)$ exists and has been determined. The existence of such a solution is a global problem. Local stability theory can only tell us whether a driver's behavior stabilizes or becomes wilder when he deviates slightly from $v_{np}(t)$. The general solution of the *linear* equation (25) is the particular solution plus the general solution of the homogeneous equation.

To study the homogeneous equation, we must find the roots r of

$$(26) \qquad \prod_{k=1}^{n}\left(1 + \frac{re^{\tau_k r}}{\lambda_k}\right) = 0.$$

The following fact is proved by Herman et al. (1959).

> The roots of $ze^z + C = 0$ all have negative real parts if and only if $0 < C < \pi/2$. If in addition $0 < C \le 1/e$, the root with the largest real part is real.

Setting $z = \tau_k r$ and $C = \tau_k \lambda_k$ transforms each factor of (26) to the form $ze^z + C$. This proves that the motion of the nth car is stable if and only if $\tau_k \lambda_k < \pi/2 = 1.57$ for $1 \le k \le n$. For a long time interval, the root with the largest real part contributes the dominant exponential term to the solution of (24). Hence the oscillatory part is highly damped if $\tau_k \lambda_k \le 1/e = 0.368$ for $1 \le k \le n$. All the drivers in Table 1 satisfy the stability criterion, but only one of them satisfies $\tau_k \lambda_k \le 0.368$.

From the preceding discussion, we see that a slight change in speed propagates down the line of cars, traveling from one car to the next after τ_k seconds. This can be viewed as a wave moving down the line of cars. From this point of view we can ask another question related to stability: What happens to the amplitude of this wave as it propagates down the line of cars? Each car individually may be stable, but the wave may increase in amplitude as it moves, thus leading to instability. A fluid dynamics model predicts the formation of a *shock wave* of acceleration or deceleration which either dies out or builds up to a maximum amplitude as it moves along the line of cars. We do not deal with this here. For a discussion see any of the books mentioned above. See also Problem 1 for a discussion of this stability question.

Let us compare the results involving time delay with a model in which reactions are instantaneous; that is, $\tau_k \equiv 0$. Equation 26 reduces to

$$\prod_{k=1}^{n} \left(1 + \frac{r}{\lambda_k} \right) = 0.$$

The roots of this are $r = -\lambda_k$. Thus the model without time lags is always stable and nonoscillatory. As the minimum λ_k increases, the roots become more negative. This increases stability, because the general solution is a linear combination of terms like $r^m e^{-\lambda_k}$. The situation in the time delay model is just the reverse of this: Stability tends to decrease as λ_k increases. Time delays are obviously important.

Equation (24) is a rather severe specialization of (23). Let us consider (23) and see how much we have to specialize it to obtain reasonable results. We assume there are only two cars and that the lead car has a constant velocity v_0. For simplicity we drop the subscript 1.

Because we have considered only stability of autonomous systems, we must eliminate the explicit time dependence of the positions. Let us adopt the convention that distance is measured from $x_0(0)$. Instead of the absolute position x of the second car, we consider the *separation* between the two cars. It is given by $s = tv_0 - x$. Equation (23) becomes

$$s''(t + \tau) = f[-s'(t) + v_0, s'(t), s(t)].$$

At the equilibrium separation s_e we have $s'' = 0$. Hence $f(v_0, 0, s_e) = 0$. Suppose that this equation has a unique root. Expanding f about this point and neglecting terms beyond the linear ones, we have

$$D^2 e^{\tau D} u \approx (f_2 - f_1) Du + f_3 u,$$

where $u = s - s_e$ and the partial derivatives f_1, f_2, and f_3 evaluated at $(v_0, 0, s_e)$. Hence we must look at the roots of

$$r^2 e^{\tau r} = (f_2 - f_1)r + f_3.$$

This can be rewritten as

(27)
$$e^z = \frac{\alpha z + \beta}{z^2},$$

where

$$\alpha = \tau(f_2 - f_1), \qquad \beta = \tau^2 f_3, \qquad \text{and} \qquad z = \tau r.$$

Since experiments indicate that the dominant effect is due to relative velocity, it is reasonable to suppose that α is negative (not positive, since s measures *separation*). If the velocities are held fixed and the separation is increased, we can expect the driver to accelerate to close the gap. Hence, β will be negative.

The study of this model cannot be completed, because we do not know what the roots of (27) are. If we want to proceed further, we should first try to study (27) analytically. If this fails, we can turn to numerical study using a computer. In view of the data in Table 1, it is reasonable to carry out such calculations with α near $-\frac{1}{2}$. Since the effect of β is probably less than the effect of α, it is reasonable to take β to be nearer to zero than α.

Although (27) seems to be a general study of the problem, it has a severe limitation: We assumed that f is differentiable. If a model builder is not careful, he can easily let this sort of assumption slip by, since most functions are in some sense "well behaved." Our assumption that f is differentiable near equilibrium implies that, except for sign, a driver responds in the same manner to a small negative relative velocity as he does to a small positive relative velocity. Actually, the acceleration and deceleration responses may be quite different because of the driver's psychology, the design of the vehicle, or both. This has been studied by G. F. Newell (1962).

If the main difference is in reaction time, the waves of acceleration and deceleration will move down the line at different speeds. It seems reasonable that deceleration will move faster. If the lead car first accelerates and then decelerates to its original speed, the two waves will eventually cancel each other out. However, if the deceleration occurs first, the acceleration wave will lag further and further behind the deceleration wave. This may provide an explanation for some of the mysterious slowdowns that occur on freeways.

PROBLEMS

1. We want to study the amplitude of a disturbance as it moves along a line of cars. For simplicity we assume that the acceleration of the first car is proportional to $\sin(\omega t)$. This is a mathematically convenient assumption which provides nonzero acceleration with no net change in velocity. It is not as restrictive as it appears at first, because we can expand $v_0(t)$ in a *Fourier series* and, by linearity, add the solutions obtained for each term separately.

   (a) Use (25) and the fact that $v_0(t) - v_0(0)$ is the real part of $Ae^{i\omega t}$ to show that $v_n(t) - v_0(0)$ is the real part of

   $$Ae^{i\omega t} \prod_{k=1}^{n} \left(1 + \frac{i\omega e^{i\omega \tau_k}}{\lambda_k}\right)^{-1}.$$

   Hint: Induct on n.

   (b) Suppose that all drivers are the same, so that $\lambda_k = \lambda$ and $\tau_k = \tau$ for all k. Deduce that the amplitude of the disturbance decreases as n increases if and only if

   $$\left|1 + \frac{i\omega e^{i\omega \tau}}{\lambda}\right| > 1.$$

   (c) Show that the above holds for all ω if and only if it holds as $\omega \to 0$, and that this yields the condition $\lambda \tau > \frac{1}{2}$.

   (d) Discuss this result in connection with Table 1.

2. Experiments indicate that, when the separation of the vehicles varies greatly, a more accurate model is provided by replacing λ_n in (24) by $\mu_n / [x_n(t) - x_{n-1}(t)]$, where μ_n is a constant. Discuss the local stability of this model.

3. The following problems are phrased rather generally. Be as specific as you must to obtain results about stability, but try to avoid unnecessary assumptions.

(a) Model the growth of a single population. Allow a time delay due to the need to mature before being able to reproduce.

(b) Same as in (a), but this time allow a time delay due to identical life spans for all members of the population. What about accidental death?

(c) Combine (a) and (b) if possible.

(d) Consider a herbivore model with a time delay built in to allow for plant recovery and perhaps delay(s) associated with the herbivore life cycle, as in (a) and (b).

4. Discuss the problem of controlling the temperature in a room as a function of how long it takes the heating unit to respond to the thermostat. For example, forced air heaters respond quickly, while steam radiators take a fairly long time.

9.4. COMMENTS ON GLOBAL METHODS

As already remarked, I consider this topic very briefly. I hope that you will get the flavor of the subject from this short discussion so that you will have some idea of the sort of problems these tools can attack.

In the physical sciences, conservation laws play an important role. A conservation law can be associated with some systems of differential equations by introducing a quantity whose time derivative is zero. For example, if $x'' = f(x)$, define

$$(28) \qquad E(t) = \frac{(x')^2}{2} - \int_0^x f(u)\,du.$$

Then $dE/dt = [x'' - f(x)]x' = 0$, and so $E(t)$ is constant. In other words, if the force acting on an object depends only on the position of the object, we can define an *energy* E which is *conserved*.

Let f be a *restoring force*; that is, $f(x)$ and x have opposite signs. Since

$$\frac{x'^2}{2} \geq 0 \qquad \text{and} \qquad -\int_0^x f(u)\,du \geq 0,$$

it follows from (28) that both of these are bounded by E and that $E \geq 0$. Thus the speed $|x'|$ is bounded. If the integrals

$$\int_0^{+\infty} f(u)\,du \qquad \text{and} \qquad \int_{-\infty}^0 f(u)\,du$$

are infinite, the position x is also bounded.

The pendulum equation (6) has the form $\theta'' = f(\theta) - h(\omega)$, where h is a frictional force and f is a restoring force provided $-\pi < \theta < \pi$. The sign of $h(\omega)$ is the same as the sign of ω. Consider $E(t)$ defined by (28) with $x = \theta$. By the previous paragraph, we have $E(t) \geq 0$. However, $E(t)$ is not constant, since

$$E'(t) = [\theta'' - f(\theta)]\omega = -h(\omega)\omega \leq 0,$$

with equality if and only if $\omega = 0$. Thus $E(t)$ decreases toward 0 as $t \to \infty$. Mathematically we say that there is global stability. Physically we say that energy loss due to friction causes the pendulum to slow down. F. Brauer (1972) discusses the motion of a pendulum when time is allowed to enter the differential equation explicitly.

The van der Pol equation,

$$(29) \qquad u'' + (u^2 - 1)\mu u' + u = 0, \qquad \mu > 0,$$

is one of the classic limit cycle problems. It arises in the study of sustained nonlinear oscillations in vacuum tubes. You should verify that the only equilibrium point is $u = 0$ and that it is unstable and oscillatory. If we approximate $\sin\theta$ by θ in the damped pendulum model in Section 9.2, it will look very much like the van der Pol equation, but $r(\theta')$ will be replaced by the term $\mu(u^2 - 1)u'$. Intuitively, if $u^2 > 1$, this term will act like a frictional force and cause damping, while if $u^2 < 1$, it will act to increase $|u'|$. Consequently $u(t)$ approaches a limit cycle. Diagrams of the limit cycle for various values of μ are given by W. E. Boyce and R. C. DiPrima (1969, p. 418).

The Poincaré–Bendixson theorem can be used to prove the intuitive result of the last paragraph. It can be stated as follows.

THEOREM.   If there is a bounded region D in the x-y plane such that any solution to the system

$$x' = f(x, y) \qquad \text{and} \qquad y' = g(x, y)$$

that starts in the region remains in it, the region contains either a stable equilibrium point or a limit cycle.

Warning: This theorem does not generalize to three dimensions.

To apply the theorem to (29), set $x = u, y = u', f(x, y) = y$, and $g(x, y) = -x - \mu(x^2 - 1)y$. Determining a region D that satisfies the theorem is not easy. You may wish to try it.

The study of global stability and limit cycles is more relevant in the life and social sciences than the study of conserved quantities. Although limit cycles are fairly common in models having nonlinear equations, they cannot occur if all the equations are linear. Hence, extreme caution should be used in modeling an essentially nonlinear phenomenon by means of a linear approximation. This is fine for studying local behavior, but it is a dangerous practice if global results are desired.

For some biological applications of global methods see J. Cronin (1977), R. H. May (1973) and T. Pavlidis (1973). For some economic applications see G. Gandolfo (1971, pp. 375–385, 421–465). Although mathematical psychology seems to be a fertile field for such methods, I am not aware of any such applications.

PROBLEM

1. We return to the predator–prey model in Section 9.2. See the discussion there. We do not wish to assume all of (13). Which do you think are the weakest assumptions? Set up some reasonable conditions to ensure that for some point (x^*, y^*) on $r(x, y) = 0$ the region

$$D = \{(x, y) | 0 \leq x \leq x^* \quad \text{and} \quad 0 \leq y \leq y^*\}$$

satisfies the Poincaré–Bendixson theorem. Draw conclusions from this.

CHAPTER 10

STOCHASTIC MODELS

You may wish to refer to the Appendix since it contains a summary of the probabilistic concepts used in this chapter. Unlike most of the earlier material, this discussion definitely requires a bit more background than 2 years of college mathematics. However, I couldn't resist the temptation to add these models, and I think they can be read with profit even if you don't fully understand the mathematics.

Radioactive Decay

The basic premise of the elementary theory of radioactive decay is that atoms have no "memory"; that is, the probability that an atom will decay during a given time interval depends only on the length of the interval and the number of neutrons and protons in the atom. In some situations, such as a chain reaction, an atom changes by absorbing a particle given off by another atom. When this doesn't happen, the decay of one atom does not affect the surrounding atoms. We consider only this case. It follows that the average rate of decay at time t is proportional to $N(t)$, the total number of undecayed atoms remaining. When $N(t)$ is large, it is reasonable to expect that most radioactive samples behave pretty much like the average. This leads to the deterministic model $N'(t) = -rN(t)$, where r is the rate of decay. The solution to this equation is

$$(1) \qquad N(t) = N_0 e^{-rt}.$$

This is fine as an approximation when the number of atoms is large, but when N_0 is small, the predictions of (1) are nonsense. For example, if $N_0 = 5$, when $t = 2/r$ we have $N(t) = 5/e^2 \approx \frac{2}{3}$. Two-thirds of an atom is nonsense. Can we construct a model that doesn't yield such nonsense?

Consider a single atom. Let T be a random variable equal to the length of time we must wait for the atom to decay. The basic assumption that an atom has no memory means that, if we have waited x minutes and the atom has not decayed, our estimate of how much longer we must wait is the same as if we had just started to observe. In mathematical language this can be written

$$\Pr\{T \geq t + x \,|\, T \geq x\} = \Pr\{T \geq t\}.$$

If $G = 1 - F$, where F is the distribution function for T, we can rewrite this as $G(t + x)/G(x) = G(t)$. In other words, $G(t + x) = G(x)G(t)$. It is well known that this implies that $G(t)$ is $e^{-\lambda t}$ for some $\lambda > 0$. We prove this under the assumption that F is differentiable at 0. The derivative of $G(t)$ is

$$G'(t) = \lim_{x \to 0} \frac{G(t + x) - G(t)}{x} = G(t)G'(0),$$

since $G(t + x) = G(x)G(t)$ and $G(0) = 1 - F(0) = 1$. With $\lambda = -G'(0)$, we obtain the desired result. The distribution function $F(t) = 1 - e^{-\lambda t}$ is called the *exponential distribution* and is associated with "memoryless" situations.

The probability that an atom has *not* decayed by time t is just $1 - F(t) = G(t)$, which is (1) with $N_0 = 1$. This is not surprising. Since $G(t)$ is the probability that any given atom has not decayed by time t, $N_0 G(t)$ is the *expected number* of undecayed atoms at time t. Thus λ is the decay rate, and (1) is just the average path of the decay process.

How closely is the average path followed? Associate with the ith atom a random variable $Y_i = Y_i(t)$ which is 1 if the atom is undecayed at time t and 0 otherwise. Then $\Pr\{Y_i = 1\} = G(t)$. The Y_i are independent by our assumption that the decay rate of an atom is independent of its surroundings. Hence the random variable

$$Y = Y_1 + Y_2 + \cdots + Y_{N_0}$$

has mean μ and variance σ^2, where

(2)
$$\begin{aligned} \mu &= \mu_1 + \mu_2 + \cdots = N_0 G(t) = N_0 e^{-\lambda t}, \\ \sigma^2 &= \sigma_1^2 + \sigma_2^2 + \cdots = N_0 G(t)[1 - G(t)]. \end{aligned}$$

Since σ provides a measure of typical deviation from the mean, σ/μ gives a measure of the typical percentage error involved in using (1). It is known as the *coefficient of variation*. By (2),

$$\frac{\sigma}{\mu} = \sqrt{\frac{1 - G(t)}{N_0 G(t)}},$$

which is small provided $N_0 G(t)$, the expected number of undecayed atoms at time t, is large. A gram of matter contains more than 10^{21} atoms, so (1) is usually a very good approximation.

There are some cases in which the coefficient of variation σ/μ may be significant. When a new radioactive isotope is produced in a particle accelerator, the number of atoms may be relatively small. This causes problems in estimating λ. In population biology, growth models like (1) are used. The population size N_0 may sometimes be sufficiently small for random fluctuations to be important.

Is it possible to obtain an exponential decay curve when we have a mixture of atoms with different decay rates? The answer to this question is no. Suppose we start out with a mixture of things that are decaying exponentially at various rates λ. If $F(l)$ is the fraction of the original mixture with $\lambda \le l$, the expected amount of the mixture undecayed at time t is

$$\mu(t) = \int e^{-\lambda t} \, dF(\lambda),$$

which can be shown to have the form e^{-kt} if and only if $F(l)$ equals 0 for $l < k$ and 1 for $l \ge k$. You may wish to try it.

Optimal Facility Location

Suppose you are faced with the problem of finding the best locations for certain facilities. To be specific and simple, consider fire stations in a large, uniform city with rectangular blocks. How can you measure the relative merit of a siting plan? How can you find the one that is best or close to best? This model is adapted from R. C. Larson and K. A. Stevenson (1972).

Suppose t is the travel time between a station and a fire. We assume that, as t increases, the situation deteriorates. Thus if siting plan A locates stations so that every point can be reached at least as quickly as in siting plan B, then A is at least as good as B. What happens if some points take longer to reach under A and others take longer under B? Various possibilities exist; for example, we could compare the average travel times or the average of the square of the travel times. We assume there exists some function $u(t)$ called the utility, and that we want to maximize the average of $u(t)$ over all points in the city. [Utility theory is discussed in many books. I recommend R. D. Luce and H. Raiffa (1958). For the two cases just mentioned we could take $u(t) = -t$ and $u(t) = -t^2$.] What do you think of this assumption? If you don't like it, can you suggest a useful alternative? Since we're going to assume $u(t)$, think about the question of what $u(t)$ should be.

By the assumption of uniformity of the city, t is roughly a linear function of travel distance s. (It is only roughly a function, because turning corners may slow the trucks down.) Let's write $f(s) = -u(t)$. Then we wish to minimize the average value of f.

Let's assume that the city streets form a rectangular grid and set up coordinate axes parallel to the streets. The travel distance between (x_1, y_1) and (x_2, y_2) is

$$|x_1 - x_2| + |y_1 - y_2|.$$

Prove this.

Suppose that there are n stations and that the city area is nA. The optimal solution is to divide the city into n equal diamonds with a station at the center of each. Of course, the geometry of the city may prevent this, in which case the best siting won't be as good as the estimate we're working out. If the area of each of the diamonds is A, the region of such a diamond is given by

$$D = \left\{(x, y): |x| + |y| \leq \sqrt{\frac{A}{2}}\right\},$$

and the average value of f is given by

(3) $$A^{-1} \iint_D f(|x| + |y|) \, dx \, dy = A^{-1} \int_0^{\sqrt{A/2}} f(s)4s \, ds.$$

Now suppose that the stations are distributed at random. Since we should easily be able to do better than random, this gives us an upper bound on what the average value of f is. If this is close to (3), we can conclude that laborious attempts at optimization will be practically useless; but if it differs considerably from (3), we can conclude that care needs to be taken in siting the stations. We must compute the expected value of the average value of f.

What is the probability that the distance between a random point in the city and the nearest station is at most s? This is the same as the probability that a station will lie in the diamond-shaped region of area $\alpha = 2s^2$ surrounding the point. If a station is placed at random, it will lie outside a region of area α with probability $1 - \alpha/nA$, since the total area of the city is nA. Thus the probability that no station will lie in the region is

$$\left(1 - \frac{\alpha}{nA}\right)^n \approx e^{-\alpha/A}.$$

It follows that the probability that the closest station will lie at a distance between s and $s + ds$ is

$$\frac{d[1 - (1 - \alpha/nA)^n]}{ds} \approx \frac{d(1 - e^{-\alpha/A})}{ds} \, ds = e^{-2s^2/A} \frac{4s}{A} \, ds,$$

since $\alpha = 2s^2$. We should average $f(s)$ times the probability over the entire area of the city. This leads to an integral. Unfortunately, the approximation we have just given is poor when α is a significant fraction of the total area of the city. If $f(s)$ does not grow exponentially with s, it will not matter because the integrand will be small. The analog to (3) is approximately

$$(4) \qquad A^{-1} \int_0^\infty f(s) e^{-2s^2/A} 4s \, ds,$$

provided the integrand becomes insignificant when $2s^2$ approaches the total area of the city. This condition is satisfied for the f we consider, provided n is greater than about 5. (You should check this out when we are studying a particular f.)

A partial check on our mathematics to date is provided by the fact that (3) and (4) both have the value c when $f(s) = c$.

Suppose we wish to minimize the average travel time. We set $f(s) = vs$, where v is velocity. Actually, travel time grows slower than linearly with s over much of the range of s for fire engines in New York City (P. Kolesar, 1975; P. Kolesar et al., 1975). Since the travel distances for random siting tend to be longer than for the best siting, it follows that the ratio between random siting and best siting travel averages will be less than what we obtain.

From (3) we have

$$(5) \qquad A^{-1} \int_0^{\sqrt{A/2}} 4vs^2 \, ds = \frac{v\sqrt{2A}}{3},$$

and from (4),

$$(6) \qquad A^{-1} \int_0^\infty e^{-2s^2/A} 4vs^2 \, ds = v \int_0^\infty e^{-2s^2/A} \, ds = v\sqrt{\frac{\pi A}{8}},$$

where the first equality is due to integration by parts and the second to the formula

$$\int_0^\infty e^{-rx^2} \, dx = \sqrt{\frac{\pi}{4r}}.$$

The ratio of (5) to (6) is $3\sqrt{\pi}/4 = 1.329$; that is, a random siting is about one-third worse than the best possible siting.

We can try other functions for $f(s)$. By the discussion of the forest fire model in Section 4.1, it seems reasonable to assume that f is a quadratic function of time with nonnegative coefficients. Hence $f(s) = as^2 + bs + c$, with $a, b, c \geq 0$, gives an upper bound. We obtain from (3) and (4), respectively,

$$\frac{aA}{4} + \frac{b\sqrt{2A}}{3} + c \qquad \text{and} \qquad \frac{aA}{2} + \frac{b\sqrt{\pi A}}{8} + c.$$

You should be able to fill in the details. The largest value of the ratio occurs when $b = c = 0$. The ratio then equals 2. Thus careful siting is more significant for a quadratic f than for a linear f.

The above results suggest that the siting of fire stations cannot be improved very much over a quick commonsense siting. Since (3) provides a lower bound on what can be achieved, any given siting can always be checked against the ideal fairly easily.

Perhaps you have already raised the objection that for something like fire fighting any improvement in siting is important. I agree, but remember that we are using a model based on an idealized city, so our results are only *approximate*. Hence the best siting for an idealized city is probably not the best siting for a real city. We can expect the two to be close but, if two site plans I and II are such that I is a bit better than II in the ideal city. It may well be that II is a bit better than I in the real city. You may wish to think about this a bit more: How can the model be made more realistic? What data should you collect to help decide where fire stations should be located in a real city? How would you go about determining sites? How is this affected by the fact that many fire stations already exist?

Distribution of Particle Sizes

If you observe the size distribution of particles in clay, material ground in a mill, or pebbles on a beach, you will probably notice that the distributions tend to have a similar shape. This suggests the existence of a common underlying principle. I would like to know what it is, so I'll make a proposal, model it, and test the model against the data. A successful model won't prove my proposal, but at least it will make it seem more likely.

It seems reasonable to suppose that particle size has been determined by a large number of small random events. Because of the central limit theorem, it is natural to look for a normal distribution. Unfortunately, the distribution or particle sizes tends to be skewed and so cannot be normal. Two main distribution laws have been proposed:

(7)
$$\text{Log normal law: } F(x) \propto \int_{-\infty}^{\log x} e^{-(t-\mu)^2/2\sigma^2} \, dt$$

$$\text{Rosin's law: } 1 - F(x) \propto e^{-rx^n}.$$

We discuss the log normal law here. The log normal distribution is discussed by J. Aitchison and J. A. C. Brown (1963) and applied by them to a variety of economic problems. The derivation given below is similar to B. Epstein's (1947). A more recent discussion of the particle size problem, with references to the literature, is given by G. V. Middleton (1970).

People have tried to fit other curves to a variety of size data, for example, the relative biomass of various species in a region, relative sizes of cities, and sizes of words. (The *Biomass* is the weight of the organisms.) Since these data are discrete, they are usually rearranged so that the items are in order of size. We then seek a model that predicts the (relative) size of the nth item in the list. See J. E. Cohen (1966), B. Mandelbrot (1965), and H. Simon (1955) for examples.

The size distribution of particles is assumed to be the result of many small changes which we will call (*breakage*) *events*. An example of an event is a wave hitting the shore. Nothing may happen during the event, or several particles may be broken and abraded. This is such a general framework that we can say very little about it.

In probability theory the basic tools for handling a long sequence of random events are *limit theorems*. We would like to use a limit theorem here if possible. To apply such theorems it is necessary to know (1) that no single event has a big effect, (2) that the events are more or less independent, and (3) that the events combine in a simple fashion.

The first condition certainly seems reasonable when averaged over all particles.

What about the second condition? Independence is closely related to the idea that knowing the past history of a particle is of no help in predicting what will happen to it. If the particles are made of two very different materials like wood and glass, this is not likely to be true. If the material is all fairly similar, this seems to be a fairly reasonable assumption. We assume that the material is all fairly similar.

The third condition is rather vague. I don't see any way to sharpen it without saying more than should be stated. What we do now is try to describe the erosion procedure and see where it leads us. Let $N_k(x)$ be the number of particles of size at most x after the kth random breakage event. We haven't yet said what we mean by "size." It could be volume, weight, a characteristic linear dimension, and so on. Let's postpone making a choice until it is useful to do so.

In the example on page 202, we saw that, when we dealt with a large number of particles, the number of undecayed particles was close to the average number. Although this sort of behavior is quite common, it is often hard to prove that it is occurring in some particular case. It seems reasonable to suppose that it holds in the present situation, but it does not seem easy to prove; therefore we simply assume that it is true. Let the average number of particles of size at most x be $M_k(x) = E(N_k(x))$. We study this as if it were an exact distribution.

Let $B_k(y\,|\,x)$ be the average number of particles of size at most y that we expect to obtain from a particle of size x during the kth breakage event. It follows from our independence assumption that B_k does not depend on any property of an individual particle except its present state. We assume that the size of the particle contains all the information about the state of a random particle relevant to breakage. Of course, this is not correct, since a long, thin particle is more likely to break than a round particle of the same size. Hence this is really an assumption that we can just look at the average behavior of particles of a given size. It is easy to show that

$$(8) \qquad M_k(y) = \int_0^\infty B_k(y \,|\, x) M'_{k-1}(x) \, dx.$$

As it stands, (8) is too general for us to try to apply a limit theorem. The following is the key assumption: *The breakage event is independent of scale.* This means that $B_k(y \,|\, x)$ depends only on the ratio y/x. This is not always a reasonable assumption. Many breakage events tend to favor the breakage of larger particles. In crushing, smaller particles are protected because their larger neighbors bear the brunt of the crushing. If particles are broken by some sort of throwing action, a scale argument shows that the smaller particles are less likely to break: The strength of a rock tends to vary with its cross section. The energy expended on a rock varies either with the cross section or with the weight, depending on the situation. If it varies with weight, energy or strength increases with size, and so larger rocks are more likely to break. These arguments indicate that our model may tend to overestimate the number of large particles.

Setting $B_k(y \,|\, x) = C_k(y/x)$, we can rewrite (8) in the form

$$(9) \qquad M_k(y) = \int_0^\infty C_k\!\left(\frac{y}{x}\right) M'_{k-1}(x) \, dx.$$

Let X_k and Y_k be random variables with distribution functions proportional to M_k and C_k, respectively. If (9) is normalized by dividing both sides by $M_k(\infty)$, the result is the formula for the distribution function of the product of two independent random variables. Hence $X_k = Y_k X_{k-1}$ which leads to

$$X_k = Y_k Y_{k-1} \cdots Y_2 Y_1 X_0.$$

Since the Y are independent and no single event has a large effect, it follows from the central limit theorem that $\log X_k$ tends to be normally distributed for large k. Thus

$$\Pr\{X_k \le x\} = \Pr\{\log X_k \le \log x\}$$

$$(10) \qquad \approx \frac{1}{\sigma\sqrt{2\pi}} \int_{-\infty}^{\log x} e^{-(t-\mu)^2/2\sigma^2} \, dt.$$

The parameters μ and σ^2 are the mean and variance of $\log X_k$, not X_k.

Now let's return to the problem of what we mean by "size." It doesn't matter whether we mean weight, a linear dimension, or a similar measure, because all powers of a log normally distributed random variable are also log normally distributed. Let's prove this. Suppose that X is log normally distributed with distribution function (7). Then

$$\Pr\{X^r \le y\} = \Pr\{X \le y^{1/r}\}$$

$$\propto \int_{-\infty}^{\log y/r} e^{-(x-\mu)^2/2\sigma^2}\, dx$$

$$\propto \int_{-\infty}^{\log y} e^{-(t-r\mu)^2/2(|r|\sigma)^2}\, dt,$$

where $t = rx$. We have shown that replacing X by X^r changes (μ, σ) to $(r\mu, |r|\sigma)$.

Statistics are often collected by passing particles through a sieve and tabulating the percentage *by weight* that passes through sieves with various mesh sizes. Our model describes particles of different sizes *by number*, not be weight. We must find out how to connect these results. We show that the distribution by weight is log normal if and only if the distribution by number is.

To study this, we need a formula for the moments of the log normal distribution $\Lambda(x; \mu, \sigma)$ which is defined by (10). We have

$$\int_0^y x^r \Lambda'(x; \mu, \sigma)\, dx = \int_0^y e^{r\log(x)} \Lambda'(x; \mu, \sigma)\, dx$$

$$= e^{r\mu + (r\sigma)^2/2} \int_0^y \Lambda'(x; \mu + r\sigma^2, \sigma)\, dx,$$

since

$$rt - \frac{(t-\mu)^2}{2\sigma^2} = r\mu + \frac{(r\sigma)^2}{2} - \frac{[t-(\mu + r\sigma^2)]^2}{2\sigma^2},$$

where $t = \log x$. We can state the above result more compactly in the form

$$(11) \qquad \int_0^y x^r \Lambda'(x; \mu, \sigma)\, dx = e^{r\mu + (r\sigma)^2/2}\Lambda(y; \mu + r\sigma^2, \sigma).$$

With $y = \infty$ and $r = 1, 2$ the mean and variance of the log normal distribution can be obtained from (11):

$$\text{Mean} = \alpha = e^{\mu + \sigma^2/2},$$
$$\text{Variance} = \beta = \alpha^2(e^{\sigma^2} - 1).$$

We are now in a position to compare distribution by weight and distribution by number. Suppose $\Lambda(y; \mu, \sigma)$ describes the distribution by number. If the particles are all roughly the same shape, setting $r = 3$ in (11), we obtain a function proportional to the distribution by weight. Hence, *distribution by weight is log normal* with the distribution function $\Lambda(x; \mu + 3\sigma^2, \sigma)$. Let α_w and β_w be the mean and the variance of this distribution. We can easily express the mean α_n of the distribution by number using α_w and β_w:

$$\alpha_n = e^{\mu + \sigma^2/2}$$
$$= e^{(\mu + 3\sigma^2) + \sigma^2/2} e^{-3\sigma^2}$$
$$= \alpha_w \left(1 + \frac{\beta_w}{\alpha_w^2}\right)^{-3}$$

How does the model fit the real world? It fits some data remarkably well and fails at other times. The data in Table 1 is taken from G. Herdan (1953, p. 130) who in turn took it from an article by S. Berg in a Danish journal. The percentage by weight of clay particles not exceeding a certain size (measured in micrometers) was tabulated. The plot on log probability paper should be a straight line. Using a least squares fit we obtain $\mu = -0.377$ and $\sigma = 1.47$ when the logarithms in (7) are taken to the base e. The third column in Table 1 shows that the fit is very good.

Table 1 Distribution of Clay Particle Sizes by Weight

Size ≤ (micrometers)	Percent	
	True	Fitted
0.106	10.0	10.2
0.147	14.9	14.7
0.25	24.6	24.6
0.38	36.4	34.4
0.65	48.3	48.5
0.96	57.5	59.0
1.41	67.6	68.8
2.15	77.5	78.1
3.25	87.3	85.5

Source: G. Herdan (1953, p. 130).

Table 2   Distribution of Sand Grain
Sizes by Weight

Size ≤ (millimeters)	Percent	
	True	Fitted
0.074	3.1	1.0
0.104	5.8	4.5
0.147	12.9	14.2
0.208	28.5	32.9
0.295	56.1	57.6
0.417	79.6	79.4
0.589	94.1	92.6
0.833	99.5	98.1
1.17	99.93	99.6

Source: G. H. Otto (1939).

The material in Table 2 was taken from G. H. Otto (1939), who obtained it by studying a sand dune in Palm Springs, California. In this case $\mu = -1.33$ and $\sigma = 0.551$. The fit is not quite as good

If you are interested in more data, you might try the article by G. M. Friedman (1958). I have not checked to see how well his data can be described by a log normal distribution.

PROBLEMS

1. When steel tapes are used to measure distance, alignment can be a problem. For example, suppose we use a 100 foot long steel tape to measure the distance between two points about a $\frac{1}{4}$ mile apart. It is unlikely that we will be able to measure along a straight line connecting the points; instead we will probably zigzag slightly. As a result, the measured distance will exceed the actual distance. The following model of the situation was adapted from B. Noble (1971, Sec. 13.6).

   (a) Suppose that the error in aligning the kth usage of a tape of length L is e_k (Figure 1). Show that, if the distance between the two points in question is about nL, the distance is overestimated by approximately

$$\delta = \frac{1}{2L} \sum_{k=1}^{n} e_k^2.$$

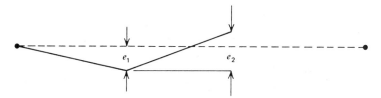

Figure 1   Errors in aligning a measuring tape.

(b) What reasonable assumptions can you make about the distribution of e_k to obtain information about the distribution of δ? Can you apply the central limit theorem?

2. The following problem was adapted from B. Noble (1971, Sec. 15.3). Suppose you are asked to decide whether or not to install a traffic signal at a pedestrian crosswalk. To arrive at an answer you need to know how long a person can expect to wait before a gap in the traffic provides enough time to cross. The only data you can expect to obtain are physical information about the street and the rate of traffic flow in cars per hour. How can this be used? For simplicity, we assume that for most of the problem the traffic all moves in the same direction.

(a) It has been found experimentally that the process of car arrival at a given point on a road can be approximated fairly well by a memoryless (Poisson) process. Show that, if the average number of cars passing the point per unit time is λ, the probability that no cars will pass during a given interval of length t is $p = e^{-\lambda t}$.

(b) Show that the expected waiting time for a gap of size at least t is roughly t/p. Is this estimate high or low? How accurate is it?

(c) Children walk at a rate of about 3.5 feet per second. If we wish the expected waiting time for a child to be at most 1 minute, obtain an estimate for the maximum permissible flow rate $\lambda_{\max}$ in cars per hour as a function of street width D. Noble gives

$$\lambda_{\max} = \frac{29{,}000(2.322 - \log_{10} D)}{D}$$

which has been adopted by the Joint Committee of the Institute of Traffic Engineers and the International Association of Chiefs of Police.

(d) How accurate is the estimate in (c)? What if the assumption in (a) is incorrect because of saturation of the roadway or because of the presence of traffic signals up the road?

(e) Discuss the situation in which traffic is moving in both directions.

3. How far apart can we expect the ends of a randomly thrown string to fall? J. L. Synge (1970) presents an interesting discussion of unsuccessful attempts to model this situation. The following is adapted from L. E. Clarke (1971) who wrote an article in response to Synge's.

   We assume that the string is made of n small stiff pieces of length l, where the angle between adjacent pieces is a random variable depending on l. Then we allow $l \to 0$. Let the location of one end of the string be the origin and let the farther end of the kth segment be at the point

$$\left(\sum_{i=1}^{k} X_i, \ \sum_{i=1}^{k} Y_i \right).$$

Let $P_i = (X_i, Y_i)$ and give it a physical interpretation.

(a) Show that the expected value of the square of the distance between the ends of the string equals

$$E(S^2) = nl^2 + 2 \sum_{i<j} E(X_i X_j + Y_i Y_j).$$

(b) Argue that we can assume

$$E(P_{i+1}|P_i, P_{i-1}, \ldots) = E(P_{i+1}|P_i),$$

and also that

$$E(P_{i+1}|P_i = (l, 0)) = (ql, 0),$$

for some $q = q(l) < 1$. Show that it is reasonable to suppose that $q(0) = 1$ and that $m = -q'(0) > 0$ is a measure of flexibility. (You should picture what is happening: We are considering shorter and shorter lengths of string of some fixed thickness.) Is it reasonable to assume that $q'(0)$ exists as we have just tacitly done?

(c) Show that

$$E(P_{i+1}|P_i) = qP_i.$$

(d) Show that, for $i < j$,

$$E(P_j|P_i) = q^{j-i}P_i,$$

and that

$$E(X_i X_j + Y_i Y_j) = E(q^{j-i}X_i^2 + q^{j-1}Y_i^2) = q^{j-i}l^2.$$

Combine this with (a) to obtain

$$E(S^2) = l^2 n + \frac{2l^2 q(q^n - nq + n - 1)}{(1 - q)^2}.$$

Hint:

$$\sum_{i<j} q^{j-i} = \sum_{i \le n} (n-i)q^i \quad \text{and} \quad \sum iq^i = q\frac{d(\sum q^i)}{dq}.$$

(e) Fix the value of $r = nl$, the length of the string, and let $l \to 0$. Show that $E(S^2/r^2) = g(mr)$, where

$$g(t) = \frac{2(e^{-t} + t - 1)}{t^2}.$$

(f) Does the result appear reasonable? I recommend that the class design and carry out an experiment to test the model. How difficult is it to throw a string at random? Do you have problems with the string tending to stick to itself? With centrifugal force when the string is thrown?

These sorts of models are closely related to random walks. The result in (d) is applicable to the problem of determining the lengths of long chain polymers. See C. Tanford (1961, Sec. 9).

4. (a) You are the manager of a delicatessen. Certain items that you stock are highly perishable. The pastries you buy from the wholesale bakery must be ordered 1 day ahead and can be kept only 1 day. How should you determine the size of your order?

 (b) Your competitor has less stringent standards than you, so he keeps pastries for 2 days. What is his optimal ordering policy? If you both have the same costs and wish to make the same profit, how will your prices compare? Will they differ substantially? How is your answer affected by the volume of business you and your competitor do?

 Note: you must make a variety of assumptions to do this problem. Discuss them.

5. This problem is adapted from H. M. Finucan (1976). Sometimes we must choose a variable x which is stochastically related to another variable y. Penalties for $y > y_0$ and $y < y_0$ may be substantially different. For example, suppose you wish to jump across a stream. Let x be the amount of effort used, y the distance of your jump, and y_0 the width of the stream. In a plant with automatic packaging, x may be the length of time a chute filling a container is open, y the weight of the product entering the container, and y_0 the minimum acceptable weight. Here we model a situation that is different from these two. When steel beams are made by continuous hot-rolling, they are cut twice. The first cut is a

rough cut as the beam emerges from the rollers. The second is a precise cut of the cool beam. The length y of the cooled rough cut beam is approximately normally distributed with mean x and variance S^2. The machinery is calibrated in terms of x. S^2 is measurable and cannot be changed except by changing the mill machinery and/or operating procedures; therefore we consider it fixed and known. If the length of the cool beam exceeds y_0, it is cut to the length y_0; if the length is less than y_0, it is rejected.

(a) Define

$$e(z) = \frac{e^{-z^2/2}}{\sqrt{2\pi}} \quad \text{and} \quad E(z) = \int_z^\infty e(t)\, dt.$$

Show that

$$P(x) = \Pr\{y \geq y_0\} = E\!\left(\frac{y_0 - x}{S}\right),$$

and that the average length of cold steel needed to produce one beam is $W(x) = x/P(x)$.

(b) Conclude that the extreme values of W are given by the solutions to

$$\frac{y_0}{S} - z = \frac{E(z)}{e(z)},$$

where $x = y_0 - Sz$. Describe a procedure for computing the value of x that *minimizes* $W(x)$. Finucan cites $y_0 = 30$ feet and $S = 2$ feet as a typical example. Show that the optimal value for x is 33 feet 11 inches. (Use a table of $e(z)$ and $E(z)$, or a table of $E(z)/e(z)$ if you have one.)

(c) Suppose undersized beams can be cut to length u_0 and used. Assume that $y_0 - u_0$ is much larger than S. Discuss a model.

(d) Can you suggest improvements in the model? Other applications? Develop a model for the packaging example cited at the beginning of the problem.

APPENDIX
SOME PROBABILISTIC
BACKGROUND

This appendix contains a hasty survey of the probability theory needed in the text. It can be used as a review for those who have had some probability theory. For those who have not had any, it can be used as an adjunct to lectures on the subject.

A.1. THE NOTION OF PROBABILITY

If I toss a fair coin, what are the chances that it will come up heads? We expect to see 50% heads in the long run and so write

$$\Pr \{\text{heads}\} = \tfrac{1}{2}.$$

This is read, "The probability of the event 'the coin lands heads up after this toss' equals $\tfrac{1}{2}$"; however, we shorten it to, "The probability of heads equals $\tfrac{1}{2}$."

What happens when we don't know the probability from a priori considerations? For example, what is the probability that a newborn baby will be a boy? We need to say very carefully what we mean. The fraction of newborn children who have been males in recent years has been 0.514 in the United States. Therefore we could say that the probability of a male child is 0.514 if the expectant mother is American. However, if you told me that she is a black American, I would recommend changing the probability to 0.506, since this is the observed fraction when the mother is a black

American. What's going on? The population I'm looking at has changed from all babies recently born to American women to all babies recently born to black American women. Note that both these populations are drawn from the past; as in all of science I'm assuming that the future will resemble the past. Although these considerations are essential for applications, they should not enter into the theoretical framework of probability theory to which we now turn our attention. The problem of estimating probabilities, to which I've alluded above, comes up again in the last paragraph of Section A.5.

DEFINITION.   Let $\mathscr{E}$ be a finite set and let Pr be a function from $\mathscr{E}$ to the nonnegative real numbers such that

$$\sum_{e \in \mathscr{E}} \text{Pr} \{e\} = 1.$$

(Note the braces instead of parentheses for the function.) We call $\mathscr{E}$ the *event set*, the elements of $\mathscr{E}$ the *simple events*, and Pr $\{e\}$ the *probability* of the simple event e.

As an illustration, consider tossing a fair coin twice. The outcomes can be denoted by the obvious notation HH, HT, TH, and TT. We can think of these as simple events and write

$$\mathscr{E} = \{HH, HT, TH, TT\}.$$

Also, Pr $\{e\} = \frac{1}{4}$ for each $e \in \mathscr{E}$. As another illustration, suppose that we toss the coin until a head occurs or until we have completed two tosses. Then the simple events can be denoted by 1, 2, and F—meaning a head at the first toss, a head at the second toss, and a failure to obtain a head. These correspond, respectively, to H, TH, and TT in the previous notation. We have

$$\mathscr{E} = \{1, 2, F\}, \qquad \text{Pr} \{1\} = \tfrac{1}{2}, \qquad \text{Pr} \{2\} = \text{Pr} \{F\} = \tfrac{1}{4}.$$

Note that the simple events in both examples are *mutually exclusive and exhaustive*; that is, exactly one occurs. This is the case in all interpretations of simple events.

If we had tossed a coin twice in the last example, we could think of event 1 as being the occurrence of either of the two simple events, HH and HT. We would write this as 1 = {HH, TT}. Thus we would write

$$\text{Pr} (HH, HT\} = \text{Pr} \{1\} = \tfrac{1}{2},$$

and read the left side as the probability of either HH or HT occurring. More generally,

DEFINITION.   For any subset S of $\mathscr{E}$ we define Pr $\{S\}$ to be the sum of Pr $\{e\}$ over all $e \in S$ and refer to it as the probability that a simple event in S will occur or, briefly, the probability that S will occur.

We can estimate Pr $\{S\}$ by sampling from $\mathscr{E}$ in such a way that each elementary event e is chosen with probability Pr $\{e\}$. (For the examples given above, our sampling can be accomplished by repeatedly tossing the coin.) If N_S of the elementary events in such a sample of size N lie in S, then N_S/N is an estimate for Pr $\{S\}$. This is the idea behind Monte Carlo simulation. We'll find it convenient to use the abbreviation Pr $\{\text{statement}\}$ for Pr $\{S\}$, where S is the set of all e such that the statement is true if $e \in S$ occurs and false if $e \notin S$ occurs. For example, in the two tosses of a fair coin, Pr $\{\geq 1 \text{ head}\}$ stands for the probability of the set $\{HT, TH, HH\}$.

We need two other concepts. After defining them, I'll discuss them briefly.

DEFINITION. The *conditional probability* of A given B is defined to be Pr $\{A \cap B\}/$Pr $\{B\}$ and is denoted by Pr $\{A|B\}$. The sets of events A and B are called *independent* if

$$\text{Pr } \{A \cap B\} = \text{Pr } \{A\} \text{ Pr } \{B\}.$$

Conditional probability is interpreted as the probability that $e \in A$ given that $e \in B$. We can think of this as restricting our attention to B: If we estimate probability by counting, as described earlier, we will estimate the probability that an event in B lies in A by $N_{A \cap B}/N_B$. Since this equals $(N_{A \cap B}/N)/(N_B/N)$, we see that the definition of conditional probability agrees with the notion of restricting our attention to the events in B.

We can think of independence as follows. Knowing that $e \in B$ gives no information about whether or not $e \in A$, since

$$(1) \qquad \text{Pr } \{A|B\} = \frac{\text{Pr } \{A \cap B\}}{\text{Pr } \{B\}} = \text{Pr } \{A\},$$

by the definitions of conditional probability and independence. By symmetry, the roles of A and B can be interchanged.

PROBLEMS

1. Prove that Pr $\{A \cup B\} = $ Pr $\{A\} + $ Pr $\{B\} - $ Pr $\{A \cap B\}$.

2. Two dice are thrown. All that matters is the sum of the two values. Formulate this in a probabilistic framework.

3. We are looking at U.S. coins minted in the 1960s. Our interest is in denomination, date, and mint. Discuss some things we could consider and cast them all in the appropriate terminology, assuming that a simple event corresponds to observing a single coin. To begin with, what is $\mathscr{E}$? Does it help to know the number of each type of coin that was minted? Why?

A.2. RANDOM VARIABLES

We're frequently not interested in simple events but only some real-valued function of them; for example, the number of heads in 100 tosses of a coin. A natural choice for the set of simple events is the 2^{100} possible sequences of heads and tails, but the function we wish to study takes on only 101 values—a considerable reduction from 2^{100}. The value of such a function depends on which simple event occurs, so it is a variable. Since it depends on something that is random, it is a random variable. Thus we have

DEFINITION.   A *random variable* is a real-valued function defined on $\mathscr{E}$.

It is conventional to use capital letters for random variables. Instead of the functional notation $X(e)$, one frequently writes simply X and talks about the value x of X. The function $\Pr\{X \leq x\}$ is called the (*cumulative*) *distribution function* for X and is important in discussing continuous probabilities. (See Section A.4.) By our convention regarding $\Pr\{\text{statement}\}$, it equals the sum of $\Pr\{e\}$ over all elementary events e with $X(e) \leq x$.

We are often interested in what values X is likely to take on; for example, if we toss our coin 100 times and count the number of heads, how many do we expect? How close to this estimate can we expect to be? We now introduce two important concepts relating to these questions.

DEFINITION.   The *expectation* or *expected value* of X is given by

$$E(X) = \sum_{e \in \mathscr{E}} X(e) \Pr\{e\},$$

and the *variance* of X is given by

$$\sigma^2(X) = \sum_{e \in \mathscr{E}} [X(e) - E(X)]^2 \Pr\{e\}.$$

[Note that in the definition functional notation is used correctly; i.e., X should not be replaced by $X(e)$ at any of its occurrences.]

The expectation is the average value of X. If we make lots of observations and compute the average value of X, it will approximate $E(X)$. The average value of X over a series of observations is denoted by $\overline{X}$. Since $\overline{X}$ is easily determined, we have a good way to estimate $E(X)$. Thus, if $E(X)$ completely determined $\Pr\{X = e\}$, we'd have a method for estimating whatever we wanted about X. We see examples of this later.

The variance is a measure of how much we can expect values of X to deviate from $E(X)$—the average value of $[X(e) - E(X)]^2$; that is,

$$\sigma^2(X) = E([X - E(X)]^2).$$

(This suggests that we can approximate σ^2 by $\overline{(X - \overline{X})^2}$. This is true, but a better estimate is given by this number times $n/(n - 1)$. We won't go into the reason here.) The larger the value of $\sigma^2(X)$, the more spread out the values of X tend to be. Stated another way, if the variance is small, then X is not likely to deviate far from $E(X)$. The following theorem makes this precise. The proof is left as a problem.

THEOREM. Chebyshev's inequality. Whenever $c > 0$,

$$\Pr\{|X - E(X)| > c\} \leq \frac{\sigma^2(X)}{c^2}.$$

In words, the probability that X differs from its expected value by more than c does not exceed its variance divided by c^2. Note that the theorem is useless if $c^2 < \sigma^2(X)$.

Some basic properties of expectation and variance are

$$E(X) = \sum_x x \Pr\{X = x\},$$

$$E(aX + bY) = aE(X) + bE(Y), \qquad E(a) = a,$$

(2) $$\sigma^2(X) = E(X^2) - E(X)^2,$$

$$\sigma^2(aX + b) = a^2\sigma^2(X), \qquad \sigma^2(a) = 0,$$

$$\sigma^2(X) \geq 0.$$

We prove the second and third. You do the others. We have

$$E(aX + bY) = \sum_e [aX(e) + bY(e)] \Pr\{e\}$$

$$= a \sum_e X(e) \Pr\{e\} + b \sum_e Y(e) \Pr\{e\}$$

$$= aE(X) + bE(Y)$$

and

$$E(a) = \sum_e a \Pr\{e\} = a.$$

for the third

$$\sigma^2(X) = E([X - E(X)]^2)$$
$$= E(X^2 - 2E(X)X + E(X)^2)$$
$$= E(X^2) - 2E(X)E(X) + E(X)^2$$
$$= E(X^2) - E(X)^2.$$

The notions of independence and conditionality can of course be carried over to random variables. Thus we say that X and Y are independent if

$$\Pr \{X = x \text{ and } Y = y\} = \Pr \{X = x\} \Pr \{Y = y\},$$

for all x and y. In other words, the events $X = x$ and $Y = y$ must be independent for all x and y. Hence knowing the value of X gives no information about the value of Y, and vice versa. The conditional expectation is defined by

$$(3) \qquad E(X \mid Y = y) = \sum_x x \Pr \{X = x \mid Y = y\}.$$

In other words, it is the average value of X on the set of events for which $Y(e) = y$. Although this is a function of y, it is often abbreviated $E(X \mid Y)$. Note that $E(E(X \mid Y))$ is simply $E(X)$, because $E(E(X \mid Y))$ is obtained by multiplying (3) by $\Pr \{Y = y\}$ and summing over y, which by simple manipulation reduces to $E(X)$.

The importance of independence is reflected in the following theorem.

THEOREM.   If X and Y are independent random variables,

$$(4) \qquad \begin{aligned} E(XY) &= E(X)E(Y), \\ \sigma^2(X + Y) &= \sigma^2(X) + \sigma^2(Y), \\ E(X \mid Y) &= E(X). \end{aligned}$$

We prove these. We have

$$\begin{aligned} E(XY) &= \sum_e X(e)Y(e) \Pr \{e\} \\ &= \sum_{x,y} xy \Pr \{X = x \text{ and } Y = y\} \\ &= \sum_{x,y} xy \Pr \{X = x\} \Pr \{Y = y\} \\ &= E(X)E(Y), \\ \sigma^2(X + Y) &= E((X + Y)^2) - (E(X + Y))^2 \\ &= E(X^2 + 2XY + Y^2) - [E(X) + E(Y)]^2 \\ &= \sigma^2(X) + \sigma^2(Y) + 2E(XY) - 2E(X)E(Y) \\ &= \sigma^2(X) + \sigma^2(Y), \end{aligned}$$

and, by (1),

$$\begin{aligned} E(X \mid Y) &= \sum_x x \Pr \{X = x \mid Y = y\} \\ &= \sum_x x \Pr \{X = x\} \\ &= E(X). \end{aligned}$$

PROBLEMS

1. Complete the proof of (2).

2. Prove Chebyshev's inequality by showing that
$$\sigma^2(X) \geq c^2 \, \Pr \{|X - E(X)| \geq c\}.$$

3. The notion of independence is extended to several sets by requiring that for any subcollection $A, B, \ldots, C$ of the sets
$$\Pr \{A \cap B \cap \cdots \cap C\} = \Pr \{A\} \, \Pr \{B\} \cdots \Pr \{C\}.$$

   Describe independence for several random variables and show that, if $X_1, \ldots, X_n$ are independent,
$$E\left(\prod_i X_i\right) = \prod_i E(X_i),$$
$$\sigma^2\left(\sum_i X_i\right) = \sum_i \sigma^2(X_i).$$

   What else can you say about the situation?

4. If X and Y are independent random variables with $\Pr \{X = x\} = f(x)$ and $\Pr (Y = y\} = g(y)$, show that
$$\Pr \{X + Y = z\} = \sum f(x)g(z - x),$$

   the sum ranging over all x for which $f(x) \neq 0$.

5. (a) Establish *Bayes' formula*:
$$\Pr \{A|B\} = \frac{\Pr \{A\} \, \Pr \{B|A\}}{\Pr \{B\}}.$$

   (b) Suppose that a diagnostic test has been developed that detects a particular disease 98% of the time when it is actually present and incorrectly "detects" in 5% of the time when it is not present. If 1% of the population has the disease, show that the probability an individual has the disease when the test says that he does is
$$\frac{(0.01)(0.98)}{(0.01)(0.98) + (0.99)(0.05)} = 0.14.$$

   In other words, 86% of the detections are incorrect.

A.3. BERNOULLI TRIALS

Consider an experiment made up of a repeated number of independent identical trials each having two outcomes; for example, coin tossing. These are called *Bernoulli trials*. Since Bernoulli trials are important, I'll discuss some of their basic properties.

We designate the outcomes of the trials by S and F for success and failure and let p be the probability that trial i ends in success. A typical simple event is a sequence containing some number s of successes and some number f of failures in some order. Since the trials are independent, probabilities multiply, and so $p^s(1 - p)^f$ is the probability of the simple event, given that exactly $s + f$ trials are performed.

Let S_n be a random variable equal to the number of successes in the first n trials. We want to study $\Pr\{S_n = k\}$. Let $\binom{n}{k}$, read "n choose k," denote the number of ways to choose k locations in an n long sequence. Then

$$(5) \qquad \Pr\{S_n = k\} = \binom{n}{k} p^k (1 - p)^{n-k}.$$

The numbers $\binom{n}{k}$ are the well-studied *binomial coefficients*. Their values turn out to be.

$$\binom{n}{k} = \frac{n(n - 1) \cdots (n - k + 1)}{1 \cdot 2 \cdots k}.$$

To study S_n it is convenient to introduce random variables that reflect the independence of the trials. Define random variables X_i by $X_i = 1$ if the ith trial succeeds, and $X_i = 0$ otherwise. Then, $S_n = X_1 + \cdots + X_n$, and the X_i are independent. One easily computes $E(X_i) = p$, and

$$\sigma^2(X_i) = p(1 - p)^2 + (1 - p)(0 - p)^2 = p(1 - p).$$

By Problem A.2.3, $E(S_n) = np$ and $\sigma^2(Sn) = np(1 - p)$.

How long must we wait for our first success? We have a problem here because there may be no success in the first n trials. To overcome this, we do computations with n fixed and then let $n \to \infty$. The answer is the expected value of a random variable that equals k if and only if the first success occurs on trial k. Hence we obtain

$$\sum k \Pr\{S_{k-1} = 0 \text{ and } X_k = 1\} = \sum k \Pr\{S_{k-1} = 0\} \Pr(X_k = 1)$$
$$= \sum k(1 - p)^{k-1} p$$
$$= p \frac{d}{dq} \sum q^k,$$

where the sums range from $k = 1$ to $k = n$. Evaluating the last sum and letting $n \to \infty$, we find that the expected waiting time for the first success equals $1/p$. Since the trials after the first success are independent of the trials leading up to the first success, we see that the expected waiting time for the jth success is j/p.

PROBLEMS

1. Let W_1 be a random variable equal to the number of Bernoulli trials until the first success.

   (a) Show that $\Pr \{W_1 = n\} = q^{n-1}p$.
   (b) What is $\sigma^2(W_1)$?

2. Let W_k be a random variable equal to the number of Bernoulli trials until the kth success.

   (a) Show that

   $$\Pr \{W_k = n\} = \binom{n-1}{k-1} q^{n-k} p^k.$$

   (b) What is $\sigma^2(W_k)$? *Hint*: Look at $X_1 + X_2 + \cdots + X_k$, where the X_i are independent and have the same distribution as W_1.

3. The circuitry in my hand calculator has a probability of failure equal to p per hour of use, independent of how long I have used it. How long can I expect the calculator to work before it fails?

4. In situations like that in the previous problem, circuits can be duplicated. Then failure does not occur until both copies of the circuit have failed. Let T be the time to failure.

   (a) Show that
   $$\Pr \{T = n\} = \Pr \{\max (X, Y) = n\},$$

   where X and Y are independent and identically distributed with the same distribution as W_1.
   (b) Show that $\Pr \{T = n\} = q^{2n-2}(1 - q^2)$, first by using (a) and second by expressing T as W_1 for some Bernoulli trials.

A.4. INFINITE EVENT SETS

Very often we want to allow an infinite event space. In this case it may be difficult to start out with elementary events. For example, consider the situation in which all the real numbers in the interval between 0 and 1 are equally likely to be chosen. We cannot assign a nonzero probability to any number, for we should then be obliged to assign the same probability to all numbers in the interval, and then the sum of the probabilities would be infinite. However, if each number has zero probability of being chosen, the sum of the probabilities will be zero. The way out of this difficulty is to ignore individual numbers and simply assign a probability to the event that the number chosen lies between x and y.

Thus we could start out with a definition of Pr as a function on the subsets of $\mathscr{E}$ having certain properties like $\Pr\{4\} \geq 0$, $\Pr\{\mathscr{E}\} = 1$, and $\Pr\{A \cup B\} = \Pr\{A\} + \Pr\{B\} - \Pr\{A \cap B\}$. This approach leads to complications. A simpler but limited approach is to work with random variables and use $\Pr\{X \leq x\}$ as the basic concept. This will satisfy our needs.

DEFINITION.   If $F(x)$ is a real-valued monotonic function satisfying

$$\lim_{x \to -\infty} F(x) = 0 \qquad \text{and} \qquad \lim_{x \to +\infty} F(x) = 1,$$

we call $F(x)$ the *distribution function* for the random variable X and write $\Pr\{X \leq x\} = F(x)$. If $f(x) = F'(x)$ exists, we call it the *density function* for X.

Roughly speaking, $f(x)\,dx$ is the probability that X lies between x and $x + dx$. By a suggestive abuse of terminology $f(x)\,dx$ is called the probability that $X = x$.

Consider the example

$$F(x) = \begin{array}{lll} 0 & \text{for} & x \leq 0, \\ 1 & \text{for} & x \geq 1, \\ x & \text{for} & 0 \leq x \leq 1. \end{array}$$

It follows that X lies in the interval between 0 and 1, since

$$\Pr\{X \leq 0\} = F(0) = 0,$$

and

$$\Pr\{X > 1\} = 1 - \Pr\{X \leq 1\} = 1 - F(1) = 0.$$

Furthermore, if $0 \leq x \leq y \leq 1$,

$$\Pr\{x < X \leq y\} = F(y) - F(x) = y - x.$$

Thus the probability that X lies in the interval (x, y) equals the length of the interval. We also have $f(x) = 1$. This is the *uniform distribution* on the interval $[0, 1]$ mentioned in the first paragraph of this section.

Consider the example $F(x) = b_i$ for $a_i \leq x < a_{i+1}$ and $i = 1, 2, \ldots, n$, where $a_0 = -\infty$, $a_{n+1} = +\infty$, $b_i = p_1 + \cdots + p_i$, and $b_n = 1$. If $a_i \leq x \leq y < a_{i+1}$, then $\Pr\{x < X \leq y\} = 0$. If $\delta > 0$ is small,

$$\Pr\{a_i - \delta < X \leq a_i\} = b_i - b_{i-1} = p_i.$$

Letting $\delta \to 0$, we see that, in some sense, $\Pr\{X = a_i\} = p_i$. Thus the step function F corresponds to a discrete distribution like those discussed in Section A.2.

Thus the present framework provides a generalization of the ideas introduced in Section A.2; however, to carry out the generalization we shall need some additional concepts, and the whole thing will appear rather theoretical. The main idea to keep in mind is that $\sum$ is replaced by $\int$ and $\Pr\{X = x\}$ is replaced by $f(x)\,dx$.

The analogy between sums and integrals suggests that we define

$$E(X) = \int_{-\infty}^{+\infty} xf(x)\,dx.$$

This has two drawbacks: First, we want to replace X by a function of X to obtain a more general definition (this is easy), and second, $f(x)$ may not exist (this is more serious). To begin with, we write

$$(6) \qquad E(g(X)) = \int_{-\infty}^{+\infty} g(x)f(x)\,dx.$$

Integrating by parts with $u = g$ and $dv = f\,dx$ we have

$$(7) \qquad E(g(X)) = g(x)F(x)\Big|_{-\infty}^{+\infty} - \int_{-\infty}^{+\infty} g'(x)F(x)\,dx.$$

This looks like a good definition for expectation, since f does not appear. Unfortunately the two terms in (7) may both be infinite. To avoid this problem we have

DEFINITION. The *expectation* of $g(X)$ is given by

$$E(g(X)) = \lim_{t \to +\infty} \left[g(t)F(t) - \int_{-\infty}^{t} g'(x)F(x)\,dx \right].$$

If $f(x)$ exists, this reduces to (6).

Now there is a question of consistency that we should consider. Let the random variable Y be defined by $Y = g(X)$. We ought to have $E(Y) = E(g(X))$. Is this the case? Suppose that g is monotonic increasing. We have

$$\Pr\{Y \leq y\} = \Pr\{X \leq g^{-1}(y)\} = F(g^{-1}(y)).$$

Hence

$$E(Y) = \lim_{u \to +\infty} \left[uF(g^{-1}(u)) - \int_{-\infty}^{u} F(g^{-1}(y)) \, dy \right]$$

by the definition. Setting $t = g^{-1}(u)$ and $x = g^{-1}(y)$, we have

$$E(Y) = \lim_{t \to +\infty} \left[g(t)F(t) - \int_{-\infty}^{t} g'(x)F(x) \, dx \right] = E(g(X)),$$

which is what we had hoped for.

The variance of X is defined to be $E([X - E(X)]^2)$.

We need to be able to handle more than one random variable simultaneously. Thus we introduce a function $F(x_1, \ldots, x_n)$ which is identified with

(8) $$\Pr\{X_1 \leq x_1 \text{ and } \cdots \text{ and } X_n \leq x_n\}.$$

Then $f = \partial^n F / \partial x_1 \cdots \partial x_n$. We require that $F \to 1$ as the $x_i \to +\infty$, $F \to 0$ as the $x_i \to -\infty$, and $f \geq 0$. The last condition can be phrased purely in terms of F to allow for the case in which f does not exist. For example, when $n = 1$, we require that $F(x) - F(x^*) \geq 0$ whenever $x \geq x^*$, and, when $n = 2$, we require that

$$F(x, y) - F(x^*, y) - F(x, y^*) + F(x^*, y^*) \geq 0$$

whenever $x \geq x^*$ and $y \geq y^*$. The $n = 1$ case corresponds to the statement that the integral of $f(t)$ from x^* to x is nonnegative, and the $n = 2$ case corresponds to the statement that the integral of $f(t, u)$ over the rectangle $[x^*, x] \times [y^*, y]$ is nonnegative. This can be generalized.

From the *joint distribution function* $F(x_1, \ldots, x_n)$ we can compute various *marginal distribution functions*, that is, probabilities like (8) in which one or more of the X_i have been deleted. For example, given $F(x, y)$ as the joint distribution function for X and Y, the distribution functions for X and Y are $\lim_{y \to +\infty} F(x, y)$ and $\lim_{x \to +\infty} F(x, y)$, respectively. You should be able to show that the density function for X is given by $\int_{-\infty}^{+\infty} f(x, y) \, dy$.

Of course, expectation is given by

$$E(g(X_1, \ldots, X_n)) = \int_{-\infty}^{+\infty} \cdots \int_{-\infty}^{+\infty} g(x_1, \ldots, x_n) f(x_1, \ldots, x_n) \, dx_1 \cdots dx_n,$$

which can be rephrased in terms of F by using n-fold integration by parts. Conditional expectation and independence also parallel Section A.2. For example,

$$E(X \mid Y) = \int_{-\infty}^{+\infty} xf(x, y) \, dx,$$

and we say that X and Y are independent if $f(x, y) = g(x)h(y)$ for some functions g and h. In this case we can choose g and h to be the density functions for X and Y. There is the old problem of replacing density functions by distribution functions. You may like to try doing this. (The idea for independence is to compute the probability that (X, Y) lies within a rectangle.)

We prove the linearity property of expectation given by (2) and leave it to you to show that (4) and the rest of (2) also generalize. For simplicity assume $f(x, y)$ exists. Then

$$E(aX + bY) = \iint (ax + by)f(x, y)\, dx\, dy$$

$$= a \int x \left[\int f(x, y)\, dy \right] dx + b \int y \left[\int f(x, y)\, dx \right] dy$$

$$= aE(X) + bE(Y).$$

PROBLEMS

1. Give the proofs asked for in the text.

2. If X and Y are independent random variables with density functions f and g, show that $Z = X + Y$ has density function

$$h(z) = \int_{-\infty}^{+\infty} f(x)g(z - x)\, dx.$$

3. Suppose that you are running a business in a service industry where demand fluctuates. (Examples include freight hauling and telephone repair.) Suppose that the wage rate is r dollars per hour and the overtime rate is s. You contract with employees for a total of N hours at the wage rate and fill any unsatisfied demand by paying overtime wages. Let X be a random variable equal to the number of service hours demanded.

   (a) If X has a density function $f(x)$, show that your expected wage costs are

   $$rN + s \int_{N}^{\infty} (x - N)f(x)\, dx.$$

   (b) Show that this is a minimum when N is chosen so that $\Pr\{X > N\} = r/s$.

   (c) Deduce the result in (b) without assuming that X has a density function.

A.5. THE NORMAL DISTRIBUTION

DEFINITION. The *normal distribution* with mean μ and variance σ^2 is given by

$$f(x) = \frac{1}{\sqrt{2\pi\sigma^2}} \exp\left(\frac{-(x - \mu)^2}{2\sigma^2}\right),$$

where $\exp(z) = e^z$.

You should verify the claims implicit in this definition; that is,

$$\int f(x)\, dx = 1, \qquad \int x f(x)\, dx = \mu, \qquad \int (x - \mu)^2 f(x)\, dx = \sigma^2.$$

You may need a table of integrals. For a normally distributed random variable X, the *standard deviation* σ provides a measure of deviation for μ that is more precise than Chebyshev's inequality, namely,

$$(9) \qquad \Pr\{|X - \mu| \le c\sigma\} = \sqrt{\frac{2}{\pi}} \int_0^c e^{-x^2/2}\, dx.$$

You should prove this.

The importance of the normal distribution stems from the fact that sums of random variables tend to be normally distributed. Consequently experimental errors are often roughly normally distributed, because they are the sum of many small effects. For biological traits such as size, the effects of genes seem often to be roughly multiplicative, and so the logarithm of size tends to be normally distributed within the adult population of a species.

These vague statements can be made mathematically precise. The result is known as the *central limit theorem* or, more accurately, central limit theorems, since there is more than one. We consider a simple one.

THEOREM. Suppose $X_1, X_2, \ldots$ are independent random variables. Let $S_n = X_1 + \cdots + X_n$. Suppose that

$$(10) \qquad \frac{\max_{1 \le i \le n} \sigma^2(X_i)}{\sigma^2(S_n)} \to 0$$

as $n \to \infty$. Define $Z_n = [S_n - E(S_n)]/\sigma(S_n)$ and let F_n be the distribution function for Z_n. Then for every z,

$$(11) \qquad \lim_{n \to \infty} F_n(z) = \frac{1}{\sqrt{2\pi}} \int_{-\infty}^z e^{-t^2/2}\, dt.$$

Since $\sigma^2(S_n) = \sum \sigma^2(X_i)$ by (4), assumption (10) ensures that as $n \to \infty$ no single X_i makes a significant contribution to the variance of Z_n. Conclusion (11) essentially says that Z_n tends to be normally distributed when n is large.

The Bernoulli trials of Section A.3 provide a simple illustration of the theorem. In this case the X_i are independent, identically distributed random variables. Thus $\sigma^2(S_n) = n\sigma^2(X_i)$, and so (10) holds. We have

$$Z_n = \frac{S_n - np}{\sqrt{npq}}.$$

A refinement of this result can be used to obtain asymptotic information about the binomial coefficients, because of (5).

Another important property of the normal distribution is that, if $X_1, \ldots, X_n$ are independent and normally distributed with means μ_i and variances σ_i^2, then $X_1 + \cdots + X_n$ is also normally distributed [with mean $\mu_1 + \cdots + \mu_n$ and variance $\sigma_1^2 + \cdots + \sigma_n^2$ by (2) and (4)]. It suffices to prove this for $n = 2$, since the rest follows easily by induction. By Problem A.4.2 the density function for $n = 2$ is

$$f(x) = \int_{-\infty}^{+\infty} f_1(t) f_2(x - t)\, dt$$

$$= \frac{1}{2\pi\sigma_1\sigma_2} \int_{-\infty}^{+\infty} \exp\left[-\frac{(t - \mu_1)^2}{2\sigma_1^2} - \frac{(x - t - \mu_2)^2}{2\sigma_2^2} \right] dt.$$

Using the identity

$$(At + B)^2 + (Ct + D)^2 = (A^2 + C^2)\left(t + \frac{AB + CD}{A^2 + C^2} \right)^2 + \frac{(BC - AD)^2}{A^2 + C^2}$$

with $A = 1/\sigma_1$, $B = -\mu_1/\sigma_1$, $C = 1/\sigma_2$, and $D = (\mu_2 - x)/\sigma_2$, we have

$$f(x) = \frac{1}{2\pi\sigma_1\sigma_2} \exp\left[\frac{-(BC - AD)^2}{2(A^2 + C^2)} \right] \sqrt{\frac{2\pi}{A^2 + C^2}},$$

which turns out to be the density function for a normal distribution with the correct mean and variance.

To change the subject, suppose that we wish to estimate some number m. It may be the expected value of some random variable, and our estimation procedure may be Monte Carlo simulation. It may be a physical constant and our estimation procedure may be experimental measurement. At any rate, after n trials we obtain n estimates x_i of m. It seems reasonable to take $\bar{x} = \sum x_i/n$ as an estimate for m. How accurate can we expect it to be? Suppose that the x_i are obtained from independent observations where the distribution function is F and the mean and variance are m and s^2, respectively. Let X_i be independent random variables with distribution function F. Then by (2) and (4),

$$\sigma^2\left(\frac{\sum X_i}{n}\right) = \frac{s^2}{n}.$$

By the central limit theorem, $\sum X_i/n$ is approximately normally distributed with mean m and variance s^2/n, and so by (9)

$$\Pr\left\{|\bar{x} - m| \leq \frac{cs}{\sqrt{n}}\right\} \approx \sqrt{\frac{2}{\pi}} \int_0^c e^{-x^2/2}\, dx.$$

Thus we expect our error to decrease as the square root of the number of trials. See the introductory part of Section 5.2 for further discussion.

PROBLEMS

1. Show that X is normally distributed with mean 0 variance 1 if and only if $(X + \mu)\sigma$ is normally distributed with mean μ and variance σ^2.

2. Suppose that X is normally distributed with mean μ and variance σ^2. Sketch the density function for X.

A.6. GENERATING RANDOM NUMBERS

In Section 5.2 I briefly discussed the generation of random numbers and provided a table of 3000 random digits. I'll treat the subject further here. There are two distinct approaches to automatically generating random numbers. The first is physical: A device is used to produce "noise" which is then translated into numbers. Examples include a noise tube and a pointer which is spun. The second method, which is the topic of this section, is to use a mathematical procedure to generate numbers which appear to be random. Numbers created in this way are not truly random, because they are produced in a repeatable manner. In fact, the numbers produced by such methods cycle—but the period of any decent method is so large as to present no problem. The idea is to devise a function f that maps the integers between 0 and M onto themselves and then, starting with x_0, compute $f(x_0) = x_1$, $f(x_1) = x_2$, ... Hopefully this will go through most of the integers between 0 and M in some seemingly random fashion. One can then use a function g to obtain random numbers of any desired sort. One objection to this procedure is that, if I tell you a random number x_n, then you can tell me its successors. This can be avoided by using certain digits of x_n to produce the random number and using other digits of x_n to compute x_{n+1}.

Here is a method for producing random numbers between 0 and 999 on a hand calculator. [For a discussion of this and many other methods for generating and testing random numbers, see D. Knuth (1969).] Choose any eight-digit number ending in 1, 3, 7, or 9. (Leading digits may be zeroes.) Define $f(x)$ to be the rightmost five digits of x times 963 and use the leftmost three digits of x (considering x to be an eight-digit number) as the random number. This can be simplified by replacing x by $x/10^5$:

1. Choose an eight-digit number x_0 of the form $d_1 d_2 d_3.d_4 d_5 d_6 d_7 d_8$, where d_8 is 1, 3, 7, or 9.
2. Define r_n to be the integer part of x_n and define x_{n+1} to be the fractional part of x_n multiplied by the number 963.

To illustrate, $x_0 = 0.12347$ leads to the sequence $x_1 = 118.90161$, $x_2 = 868.25043$, $x_3 = 241.16409$, and so on. The first four random numbers are 000 118 868 241.

In Section 5.2 it was pointed out that, if X is uniformly distributed on $[0, 1]$, then $Y = F^{-1}(X)$ has the distribution function F. To see this note that, since F is monotonic,

$$\Pr\{Y \le y\} = \Pr\{F(Y) \le F(y)\} = \Pr\{X \le F(y)\},$$

which equals $F(y)$, since X is uniformly distributed on $[0, 1]$.

Since F^{-1} is not easily computed for the normal distribution, a table or the central limit theorem should be used. To use the latter, simply generate a sequence of random numbers and apply the theorem to them. For example, if $X_1, \ldots, X_n$ are generated to be uniformly distributed on $[0, 1]$,

$$\left(X_1 + \cdots + X_n - \frac{n}{2}\right)\sqrt{\frac{12}{n}}$$

is approximately normally distributed with mean 0 and variance 1. A convenient and almost certainly large enough value for n is 12. Here is a simple table based on F^{-1} for the normal distribution. I recommend using it when doing calculations by hand or on a hand calculator. It is used as follows.

	0	1	2	3	4	5	6	7	8	9
0, 1	0.00	0.02	0.05	0.08	0.10	0.13	0.15	0.18	0.20	0.23
2, 3	0.25	0.28	0.31	0.33	0.36	0.39	0.41	0.44	0.47	0.50
4, 5	0.52	0.55	0.58	0.61	0.64	0.67	0.70	0.74	0.77	0.81
6, 7	0.84	0.88	0.92	0.95	0.99	1.04	1.08	1.13	1.18	1.23
8, 9	1.28	1.34	1.41	1.48	1.56	1.64	1.8	1.9	2.1	2.3

Generate two random digits Y_1 and Y_2. If $Y_1 = Y_2 = 0$, reject the pair and try again. Find Y_1 in the leftmost column and Y_2 in the top row. Read off the number X, changing its sign if Y_1 is odd. This is normally distributed with mean 0 and variance 1. Hence $(X + \mu)\sigma$ is normally distributed with mean μ and variance σ^2.

A.7. LEAST SQUARES

The racing shell model in Section 2.1 predicts a relationship of the form $h(x) = Cx^{-1/9}$ where x is the number of oarsmen and $h(x)$ is the best possible time in a race. Of course this is only approximate, since shell designs do not quite fit the model we proposed. Furthermore, we can only estimate the best possible times by using data which may be biased by such things as nonideal team performance, currents, and winds. Thus we obtain for various values of x (namely 1, 2, 4, and 8) estimates y for $h(x)$. What value of C gives the best fitting curve? How good is the exponent $-\frac{1}{9}$—what is the best fitting curve of the form Cx^m?

In general, we have a function $h(x)$ depending on certain parameters and we have estimates y_i of $h(x_i)$. We wish to determine the best values for the parameters. What should we do? To make any progress, we need to make some additional assumptions. Let's start with a simple situation and then return to the racing shell problem.

In Section 2.2 we predicted that for a perfect pendulum in a fixed gravitational field, the period is $\tau = C(\theta)\sqrt{l}$ where θ is angle of swing, l is length, and C is an unknown function. Let's test this by constructing pendulums of various lengths, starting them swinging at some fixed angle θ_0, and measuring the period. We can then plot τ versus $\sqrt{l}$ and see if we obtain a straight line. Of course, there will be errors in measuring τ and l, and in setting the angle of swing equal to θ_0. (The fact that the pendulum is not perfect can probably be neglected. See Section 9.2.) From another point of view, we are making errors in estimating $\tau(l) = C(\theta_0)\sqrt{l}$, both by measuring at the wrong point (θ_0 and l in error) and by measuring τ incorrectly. This suggests that after many repetitions with a given l we might obtain estimates $\tau(l)$ which are normally distributed about the predicted value $C(\theta_0)\sqrt{l}$. In other words, τ is normally distributed with mean $C(\theta_0)\sqrt{l}$ and unknown variance $\sigma^2(l)$. We make measurements for various l and thereby obtain pairs (l_i, τ_i) where τ_i is sampled from a normal distribution with mean $C(\theta_0)\sqrt{l_i}$ and variance $\sigma_i^2 = \sigma^2(l_i)$.

What is the best estimate for $C(\theta_0)$? We can interpret "best" to mean "estimate which maximizes the probability of being close to the observed values." Let $\varepsilon_i > 0$ be very small. The probability that a sampled value τ would be within ε_i of τ_i is

$$\frac{1}{\sqrt{2\pi}\,\sigma_i} \int_{\tau_i - \varepsilon_i}^{\tau_i + \varepsilon_i} \exp\left(-\frac{(t - C(\theta_0)\sqrt{l_i})^2}{2\sigma_i^2} \right) dt$$

$$\approx \frac{2\varepsilon_i}{\sqrt{2\pi}\,\sigma_i} \exp\left(-\frac{(\tau_i - C(\theta_0)\sqrt{l_i})^2}{2\sigma_i^2} \right).$$

If the observations are independent, we may multiply this probability for various values of i to obtain the joint probability. If the σ_i are independent of the parameter $C(\theta_0)$, this joint probability will be a maximum when

$$\sum \frac{(\tau_i - C(\theta_0)\sqrt{l_i})^2}{2\sigma_i^2}$$

is a minimum. We can find $C(\theta_0)$ to minimize this by setting $\partial \sum / \partial C(\theta_0)$ equal to zero and solving for $C(\theta_0)$. This approach is stated in general form in the following theorem.

THEOREM. Least squares. If Y_i are independent, normally distributed random variables with means $h(x_i)$ and variances σ_i^2 independent of h, then the probability of each Y_i simultaneously being within ε_i of y_i is maximized by selecting the function h for which

(12) $$\sum \frac{(y_i - h(x_i))^2}{\sigma_i^2}$$

is a minimum.

Usually the assumptions of the theorem cannot be verified (in fact, they are usually incorrect), and the variances cannot be estimated. The usual procedure is to apply the theorem anyway and assume that all the variances are equal. Thus we minimize

$$(13) \qquad \sum (y_i - h(x_i))^2$$

in most cases. This is what would be done in the pendulum problem discussed before the theorem.

Let's apply the theorem to the racing shell problem. Let Y_i be the best observed time for a shell with x_i men. Of course, we cannot hope to verify the hypotheses of the theorem or estimate σ_i. We make the usual assumption that the theorem holds and the σ_i are equal: We assume that the Y_i are independent normally distributed random variables with means $Cx_i^{-1/9}$ and equal variances. We wish to minimize (13) where $h(x_i) = Cx_i^{-1/9}$ and y_i is an observed best time. By setting the partial derivative with respect to C equal to zero we obtain

$$(14) \qquad \sum x_i^{-1/9}(Cx_i^{-1/9} - y_i) = 0.$$

This is a linear equation in C, so it is easily solved when the values of x_i and y_i are known. Instead of looking at x versus y as in (14), we can consider $\log x$ versus $\log y$ as suggested in Section 2.1. Then y_i in the theorem is the logarithm of the time, x_i is the logarithm of the number of men, and $h(x_i)$ is $\log C - x_i/9$. However, we have already set y_i equal to the time and x_i equal to the number of men. We will keep this notation rather than the notation of the theorem. Thus we wish to minimize

$$\sum \left[\log y_i - K + \frac{\log x_i}{9}\right]^2,$$

where $K = \log C$. In this case we've assumed that $\log Y_i$ is normally distributed with mean $\log C - (\log x_i)/9$ and variance independent of i. This is inconsistent with our assumptions about Y_i leading to (14). Setting the partial derivative with respect to K equal to zero we obtain

$$(15) \qquad \sum \log y_i - K + \frac{\log x_i}{9} = 0.$$

Equations (14) and (15) give different values for C (see the accompanying table). Which is correct? Probably neither one, since our assumptions about Y_i and $\log Y_i$ are assuredly wrong; however, both give fairly good fits to the data and the fits are about the same. Now suppose that we want to fit the exponent as well, that is, find the best $h(x)$ having the form Cx^{-r}. In this case, the second method is preferable. This is not for any theoretical reason, but simply because it is much easier to find the values of C and r that minimize (13) in this case. The equations for $K = \log C$ and r are

$$\sum \log y_i - K + r \log x_i = 0,$$

(16)

$$\sum \log y_i \log x_i - K \log x_i + r (\log x_i)^2 = 0.$$

The following table compares values obtained by the various methods. The data comes from Table 1 in Chapter 2. Since different races may be run under different conditions, it was not clear how I should interpret "best time." Should I do separate fits for each of the four races? A fit to the average of the best times of the four races? A fit to the overall best time? I fit the average best time and the overall best time. Once C and r have been determined using (14), (15), or (16), it is possible to compute $h(x_i)$. This I have also done. Note that the fit is fairly good, and the estimates for r via (16) support the model's prediction that $r = \frac{1}{9}$.

		Average Best Time				Overall Best Time		
		(14)	(15)	(16)		(14)	(15)	(16)
C		7.44	7.35	7.29		7.31	7.21	7.21
r		$\frac{1}{9}$	$\frac{1}{9}$	0.104		$\frac{1}{9}$	$\frac{1}{9}$	0.111
1	7.22	7.44	7.35	7.29	7.16	7.31	7.21	7.21
2	6.88	6.89	6.81	6.78	6.77	6.77	6.68	6.68
4	6.34	6.38	6.30	6.31	6.13	6.27	6.18	6.18
8	5.84	5.91	5.83	5.88	5.73	5.80	5.72	5.72

A.8. THE POISSON AND EXPONENTIAL DISTRIBUTIONS

Two closely related distributions are the Poisson, a discrete distribution given by $\Pr\{X = k\} = e^{-\lambda}\lambda^k/k!$ and the exponential, a continuous distribution given by $\Pr\{T \le t\} = 1 - e^{-vt}$. They both have mean and variance $\lambda = 1/v$. Prove it. The exponential is associated with waiting times between rare events, and the Poisson with the number of rare events in a given time interval. The following examples illustrate this.

1. Suppose we distribute $N\lambda$ items into N boxes. Let X be the number of items in the ith box. If the items are distributed independently and each box is equally likely to be chosen, $\Pr\{X = k\} \to e^{-\lambda}\lambda^k/k!$ as $N \to \infty$.

2. Suppose that in a small time interval Δt an event has probability $v \, \Delta t$ of occurring, independent of what has happened in the past. The waiting time T between two successive occurrences is exponentially distributed.

3. Closely related to this is failure of a product. If the probability of failure in the time interval Δt is $v \, \Delta t$ given that the product hasn't failed up to that time, the waiting time to failure is exponentially distributed.

4. Let's return to example 2. Let X be the number of occurrences of the event between t and $t + \tau$. Then X is Poisson distributed with $\lambda = \tau v$, where v is the parameter of the exponential distribution in example 2.

These examples merit more discussion.

We can think of example 1 as a Bernoulli trial situation. If an item is placed in the ith box, this is a success. We then have

$$\Pr \{X = k\} = \binom{N\lambda}{k} q^{N\lambda - k} p^k,$$

where $p = 1/N$. The claim in example 1 follows from

$$\binom{N\lambda}{k} N^{-k} \to \frac{\lambda^k}{k!} \quad \text{and} \quad q^{N\lambda} \to e^{-\lambda}$$

as $N \to \infty$. Hence the Poisson is a limiting case of Bernoulli trials.

The exponential is obtained similarly as a limit. In example 2, T is simply the waiting time to the first success. Consider a situation in which the time between Bernoulli trials is Δt and the probability of success is $v \, \Delta t = p$. The probability of a first success at time $\Delta t[T/\Delta t]$ is $q^{[T/\Delta t]} p$. (The square brackets here denote "largest integer not exceeding.") For small $v \, \Delta t$ this is approximately $v e^{-vT} \Delta t$. Hence $f(T) = v e^{-vT}$.

The relationship between the exponential and Poisson distributions asserted in example 4 is easily proved: In each time interval Δt, the probability of success is $p = v \, \Delta t$, so after N time intervals the probability of k successes is $\binom{N}{k} q^{n-k} p^k$. Setting $\Delta t = \tau/N$ and letting $N \to \infty$, we obtain the desired result.

REFERENCES

Numbers at the end of a reference refer to chapters and sections in which the item is mentioned. An asterisk indicates that a book is not very specialized and may be of particular interest to you.

Ackoff, R. L. and M. W. Sasieni (1968). *Fundamentals of Operations Research*. Wiley. 4.1.

Aitchison, J. and J. A. C. Brown (1963). *The Lognormal Distribution with Special Reference to Its Uses in Economics*. Cambridge University Press. 10.

Almond, J. (1965). *Proceedings of the Second International Symposium on the Theory of Road Traffic Flow, London, 1963*. The Organisation for Economic Co-operation and Development, Paris. 9.3.

Altman, P. L. and D. M. Dittmer, eds. (1964). *Biology Data Book*. Federation of American Societies for Experimental Biology. 2.1.

Armstrong, J. S. (1967). Derivative of theory by means of factor analysis or Tom Swift and his electric factor machine. *Amer. Stat.* **21**: 17–21. Reprinted in R. L. Day and L. J. Parsons, eds. (1971). *Marketing Models*. Intext: 413–421. 1.5.

Arrow, K. J. (1963). *Social Choice and Individual Values* (2nd ed.). Cowles Foundation Monograph 12. Wiley. 6.

*Ashton, W. D. (1966). *The Theory of Road Traffic Flow*. Methuen. 9.3.

Audley, R. J. (1960). A stochastic model for individual choice behavior. *Psychol. Rev.* **67**: 1–15. Reprinted in R. D. Luce, R. R. Bush, and E. Galanter, eds. (1963). *Readings in Mathematical Psychology*. Vol. 1. Wiley: 263–277. 5.1.

Bailey, N. T. J. (1976). *The Mathematical Theory of Infectious Diseases and Its Applications* (2nd ed.). Hafner. 9.2.

Barker, S. B., G. Cumming, and K. Horsfield (1973). Quantitative morphometry of the branching structure of trees. *J. Theor. Biol.* **40**: 33–43. 5.2.

Bartlett, A. A. (1973). The Frank C. Walz lecture halls: A new concept in the design of lecture auditoria. *Amer. J. Phys.* **41**: 1233–1240. 1.5.

Bartlett, M. S. (1972). Epidemics. In J. M. Tanur et al., eds. *Statistics: A Guide to the Unknown.* Holden-Day: 66–76. 8.2.

*Bass, F. M. et al., eds. (1961). *Mathematical Models and Methods in Marketing.* Irwin.

Baylis, J. (1973). The mathematics of a driving hazard. *Math. Gaz.* **57**: 23–26. 8.1.

Bender, E. A. and L. P. Neuwirth (1973). Traffic flow: Laplace transforms. *Amer. Math. Mon.* **80**: 417–423. 9.3.

*Bochner, S. (1966). *The Role of Mathematics in the Rise of Science.* Princeton University Press. 1.1.

*Boot, J. C. G. (1967). *Mathematical Reasoning in Economics and Management Science.* Prentice-Hall. 3.3.

Boyce, W. E. and R. C. DiPrima (1969). *Elementary Differential Equations and Boundary Value Problems.* Wiley. 9.4.

Brauer, F. (1972). The nonlinear simple pendulum. *Amer. Math. Mon.* **79**: 348–355. 9.4.

Brauer, F. and J. A. Nohel (1969). *Qualitative Theory of Ordinary Differential Equations.* Benjamin. 9.0.

Braun, M. (1975). *Differential Equations and Their Applications.* Springer-Verlag. 9.2.

Bridgman, P. W. (1931). *Dimensional Analysis.* Yale University Press. 2.2.

Brock, V. E. and R. H. Riffenburgh (1959). Fish schooling: A possible factor in reducing predation. *J. Cons. Perm. Int. Explor. Mer* **25**: 307–317. 6.

Brown, A. A., F. T. Hulswit, and J. D. Kettelle (1956). A study of sales operations. *Oper. Res.* **4**: 296–308. Reprinted in R. L. Day and L. J. Parsons (1971, pp. 3–16). 1.5.

Buchanan, N. S. (1939). A reconsideration of the cobweb theorem. *J. Polit. Econ.* **47**: 67–81. Reprinted in R. V. Clemence, ed. (1950). *Readings in Economic Analysis.* Vol. 1. Addison-Wesley: 46–60. 3.3.

Bush, R. R. and F. Mosteller (1959). A comparison of eight models. In R. R. Bush and W. K. Estes, eds. *Studies in Mathematical Learning Theory.* Stanford University Press: 293–307. Reprinted in P. F. Lazarsfeld and N. W. Henry (1966, pp. 335–349). 5.1.

*Carrier, G. F. (1966). *Topics in Applied Mathematics.* Vol. 1. (Notes by N. D. Fowkes.) Mathematical Association of America. 8.2.

Carson, R. L. (1961). *The Sea around Us* (rev. ed.). Oxford University Press. 2.2.

Chandler, R. E., R. Herman, and E. W. Montroll (1958). Traffic dynamics: Studies in car following. *Oper. Res.* **6**: 165–184. 9.3.

Clark, C. W. (1973). The economics of overexploitation. *Science* **181**: 630–634. 4.1.

Clark, C. W. (1976). *Mathematical Bioeconomics*. Wiley-Interscience. 4.1.

Clarke, L. E. (1971). How long is a piece of string? *Math. Gaz.* **55**: 404–407. 10.

Coale, A. J. (1971). Age patterns of marriage. *Popul. Stud. (London)* **25**: 193–214. 8.1.

Cohen, J. E. (1966). *A Model of Simple Competition*. Harvard University Press. 10.

Cohen, J. E. (1971). *Casual Groups of Monkeys and Men: Stochastic Models of Elemental Social Systems*. Harvard University Press. 5.1.

*Cohen, K. J. and R. M. Cyert (1965). *Theory of the Firm: Resource Allocation in a Market Economy*. Prentice-Hall. 3.2, 4.2.

Cohn, T. E. and D. J. Lasley (1976). Binocular vision: Two possible central interactions between signals from two eyes. *Science* **192**: 561–563. 6.

*Coleman, J. S. (1964). *Introduction to Mathematical Sociology*. Macmillan. 8.1.

Coleman, J. S. and J. James (1961). The equilibrium size distribution of freely-forming groups. *Sociometry* **24**: 36–45. 5.1.

Crank, J. (1962). *Mathematics and Industry*. Oxford University Press. 1.5.

Cronin, J. (1977). Some mathematics of biological oscillations. *SIAM Rev.* **19**: 100–138. 9.4.

Cundy, H. M. (1971). Getting it taped. I. *Math. Gaz.* **55**: 43–47. 6.

Cyert, R. M. and J. G. March (1963). *A Behavioral Theory of the Firm*. Prentice-Hall. 3.2.

Davis, D. D. (1962). Allometric relationships in lions vs. domestic cats. *Evolution* **16**: 505–514. 2.1.

Day, R. L. and L. J. Parsons, eds. (1971). *Marketing Models: Quantitative Applications*. Intext.

Douglas, J. F. (1969). *An Introduction to Dimensional Analysis*. Pitman. 2.2.

*Engineering Concepts Curriculum Project (1971). *The Man-Made World*. McGraw-Hill. A good high school text. 1.2.

Epstein, B. (1947). The mathematical description of certain breakage mechanisms leading to the logarithmico-normal distribution. *J. Franklin Inst.* **244**: 471–477. 10.

Evans, J. W., P. D. Wagner, and J. B. West (1974). Conditions for reduction of pulmonary gas transfer by ventilation-perfusion inequality. *J. Appl. Physiol.* **36**: 533–537. 6.

Ezekiel, M. (1937–1938). The cobweb theorem. *Quart. J. Economics* **52**: 255–280. Reprinted in G. Haberler, ed. (1944) *Readings in Business Cycle Theory.* Irwin. 3.3.

Fantino, E. and G. S. Reynolds (1975). *Introduction to Contemporary Psychology.* Freeman. 2.1.

Finucan, H. M. (1976). The silver anniversary of an optimization result in rolling-mill practice. *Oper. Res.* **24**: 373–377. 10.

Ford, L. P. (1955). *Differential Equations* (2nd ed.). McGraw-Hill. 9.3.

Friedman, G. M. (1958). Determination of sieve-size distribution from thin-section data for sedimentary petrological studies. *J. Geol.* **66**: 394–416. 10.

*Gandolfo, G. (1971). *Mathematical Methods and Models in Economic Dynamics.* American Elsevier. 9.2, 9.4.

Gazis, D. C. and R. B. Potts (1965). The over-saturated intersection. In J. Almond (1965, pp. 221–237). 4.2.

Gilpin, M. E. (1973). Do hares eat lynx? *Amer. Nat.* **107**: 727–730. 9.2.

Goel, N. S., S. C. Maitra, and E. W. Montroll (1971). On the Volterra and other nonlinear models of interacting populations. *Rev. Mod. Phys.* **43**: 231–276. Reprinted (1971) as a monograph under the same title. Academic Press. 9.2.

Goodman, L. A. (1961). Some possible effects of birth control on the human sex ratio. *Ann. Hum. Genet.* **25**: 75–81. Reprinted in P. A. Lazarsfeld and N. W. Henry (1968, pp. 311–317). 5.1.

Griffith, J. S. (1968). Mathematics of cellular control processes. I. Negative feedback to one gene. *J. Theor. Biol.* **20**: 202–208. 8.2.

Hadley, G. and T. M. Whitin (1963). *Analysis of Inventory Systems.* Prentice-Hall. 4.1.

*Haight, F. A. (1963). *Mathematical Models for Traffic Flow.* Academic Press. 9.3.

Haldane, J. B. S. (1928). On being the right size. In J. B. S. Haldane, ed. *Possible Worlds.* Harper. Reprinted in J. R. Newman (1956, pp. 952–957). 2.1.

Hammersley, J. M. (1961). On the statistical loss of long-period comets from the solar system. II. In J. Neyman (1961, pp. 17–78). 5.2.

Henmon, V. A. C. (1911). The relation of the time of a judgment to its accuracy. *Psychol. Rev.* **18**: 186–201. 5.1.

Herdan, G. (1953). *Small Particle Statistics.* Elsevier. 10.

Herman, R., E. W. Montroll, R. B. Potts, and R. W. Rothery (1959). Traffic dynamics: Analysis of stability in car following. *Oper. Res.* **7**: 86–103. 9.3.

Hernes, G. (1972). The process of entry into first marriage. *Amer. Sociol. Rev.* **37**: 173–182. 8.1.

Higgins, J. (1971). Getting it taped. II. *Math. Gaz.* **55**: 47–48. 6.

Homans, G. (1950). *The Human Group.* Harcourt Brace. 3.3.

Intriligator, M. D. (1973). Strategy and arms races, an application of ordinary differential equations to problems of national security. In P. J. Knopp and G. H. Meyer (1973, pp. 253–377 and parts of 291–298). 3.2.

Jensen, A. (1966). Safety-at-sea problems. In D. B. Hertz and J. Melese, eds. *Proceedings of the Fourth International Conference on Operational Research.* Wiley-Interscience: 362–370. 1.2.

Kármán, T. v. and M. A. Biot (1940). *Mathematical Methods in Engineering.* McGraw-Hill. 8.2, 9.2.

Keith, L. B. (1963). *Wildlife's Ten-Year Cycle.* University of Wisconsin Press. 9.2.

Kemeny, J. G. (1973). What every college president should know about mathematics. *Amer. Math. Mon.* **80**: 889–901. 5.1.

*Kemeny, J. G. and J. L. Snell (1962). *Mathematical Models in the Social Sciences.* Ginn. 9.2.

Kerr, R. H. (1961). Perturbations of cometary orbits. In J. Neyman (1961, pp. 149–152). 5.2.

Kintsch, W. (1963). A response time model for choice behavior. *Psychometrika* **28**: 27–32. 5.1.

Kleiber, M. (1961). *The Fire of Life: An Introduction to Animal Energetics.* Wiley. 2.1.

*Kline, S. J. (1965). *Similitude and Approximation Theory.* McGraw-Hill. 2.2.

Knopp, P. J. and G. H. Meyer, eds. (1973). *Proceedings of a Conference on the Application of Undergraduate Mathematics in the Engineering, Life, Managerial and Social Sciences.* Georgia Institute of Technology.

Knuth, D. E. (1969). *The Art of Computer Programming. II: Seminumerical Algorithms.* Addison-Wesley. A.6.

Kolesar, P. (1975). A model for predicting average fire engine travel times. *Oper. Res.* **33**: 603–613. 10.

Kolesar, P., W. Walker, and J. Hausner (1975). Determining the relation between fire engine travel times and travel distances in New York City. *Oper. Res.* **33**: 614–627. 10.

Kupperman, R. H. and H. A. Smith (1972). Strategies of mutual deterrence. *Science* **176**: 18–23. 5.1.

Land, K. C. (1971). Some exhaustible Poisson process models of divorce by marriage cohort. *J. Math. Sociol.* **1**: 213–232. 8.1.

Langhaar, H. L. (1951). *Dimensional Analysis and the Theory of Models.* Wiley. 2.2.

Larson, R. C. and K. A. Stevenson (1972). On insensitivities in urban redistricting and facility location. *Oper. Res.* **20**: 595–612. 10.

*Lave, C. A. and J. G. March (1975). *An Introduction to Models in the Social Sciences.* Harper and Row. 1.5.

*Lazarsfeld, P. A. and N. W. Henry, eds. (1968). *Readings in Mathematical Social Science.* M.I.T. Press.

*Leigh, E. G. Jr. (1971). *Adaptation and Diversity.* Freeman. 1.5.

Leopold, L. B., M. G. Wolman, and J. P. Miller (1964). *Fluvial Processes in Geomorphology.* Freeman. 5.2.

Levary, G. (1956). A pocket-sized case study in operations research concerning inventory markdown. *J. Oper. Soc. Amer.* **4**: 738–740. 6.

Levins, R. (1968). *Evolution in Changing Environments.* Princeton University Press. 1.2, 4.2.

Lin, C. C. and L. A. Segel (1974). *Mathematics Applied to Deterministic Problems in the Natural Sciences.* Macmillan. 1.5.

Luce, R. D. and H. Raiffa (1958). *Games and Decisions.* Wiley. 6, 10.

MacArthur, R. H. and E. O. Wilson (1967). *The Theory of Island Biogeography.* Princeton University Press. 3.2.

Mandelbrot, B. (1965). Information theory and psycholinguistics. In B. B. Wolman, ed. *Scientific Psychology.* Basic Books. Reprinted with additions in P. F. Lazarsfeld and N. W. Henry (1968, pp. 350–368). 10.

*Martin, M. J. C. and R. A. Denison (1970). *Case Exercises in Operations Research.* Wiley.

May, R. M. (1972). Limit cycles in prey-predator communities. *Science* **177**: 900–902. 9.2.

May, R. M. (1973). *Stability and Complexity in Model Ecosystems.* Princeton University Press. 9.2, 9.4.

May, R. M. (1975). Biological populations obeying difference equations: Stable points, stable cycles and chaos. *J. Theor. Biol.* **51**: 511–524. 7.4.

*Maynard Smith, J. (1968). *Mathematical Ideas in Biology*. Cambridge University Press. 2.1, 8.2.

McKelvey, R. D. (1973). Some theorems on electoral equilibrium under two-candidate competition. In P. J. Knopp and G. H. Meyer (1973, pp. 1–20). 4.2.

McMahon, T. A. (1971). Rowing: A similarity analysis. *Science* **173**: 349–351. 2.1.

McMahon, T. A. (1973). Size and shape in biology. *Science* **179**: 1201–1204. 2.1.

Metelli, F. (1974). The perception of transparency. *Sci. Amer.* **230**(4): 90–98. 6.

Middleton, G. V. (1970). Generation of the log-normal frequency distribution in sediments. In M. A. Romanova and O. V. Sarmanov, eds. *Topics in Mathematical Geology*. (Translated from the Russian.) Consultants Bureau: 34–42. 10.

Mudahar, M. S. and R. H. Day (1974). A generalized cobweb model for an agricultural sector. Mathematics Research Center Technical Summary Report 1453. University of Wisconsin, Madison. 3.3.

Murdick, R. G. (1970). *Mathematical Models in Marketing*. Intext. 4.1.

Nash, J. F., Jr. (1950). The bargaining problem. *Econometrica* **18**: 155–162. Reprinted in K. J. Arrow. ed. (1971). *Selected Readings from Econometrica*. Vol. 2. M.I.T. Press: 204–211. 6.

Neher, P. A. (1971). *Economic Growth and Development*. Wiley. 3.3.

Netter, F. H. *The Ciba Collection of Medical Illustrations*. Several volumes and dates. Ciba Pharmaceutical. 4.1.

Newell, G. F. (1962). Theories of instability in dense highway traffic. *J. Oper. Res. Soc. Jap.* **5**(1): 9–54. 9.3.

Newman, J. R., ed. (1956). *The World of Mathematics*. Vol. 2. Simon and Schuster.

Neyman, J., ed. (1961). *Proceedings of the Fourth Berkeley Symposium on Mathematical Statistics and Probability, 1960*. Vol. 3. University of California Press.

*Noble, B. (1971). *Applications of Undergraduate Mathematics in Engineering*. Macmillan. 4.2, 9.2, 10.

Norris, K. S. (1967). Color adaptation in desert reptiles and its thermal relationships. In W. W. Milstead, ed. *Lizard Ecology: A Symposium*. University of Missouri Press: 162–229. 6.

Notari, R. E. (1971). *Biopharmaceutics and Pharmacokinetics: An Introduction*. Marcel Dekker. 8.2.

Otto, G. H. (1939). A modified logarithmic probability graph for the interpretation of mechanical analyses of sediments. *J. Sediment. Pet.* **9**: 62–76. 10.

Parks, G. M. (1964). Development and application of a model for suppression of forest fires. *Manage. Sci.* **10**: 760–766. 4.1.

Pavlidis, T. (1973). *Biological Oscillators: Their Mathematical Analysis.* Academic Press. 8.2, 9.4.

*Pignataro, L. J. (1973). *Traffic Engineering Theory and Practice.* Prentice-Hall. 9.3.

Plattner, S. (1975). Rural market networks. *Sci. Amer.* **232**(5): 66–79. 5.2.

Pontryagin, L. S. (1962). *Ordinary Differential Equations.* (Translated from the Russian.) Addison-Wesley. 9.2.

Pulliam, H. R. (1973). On the advantages of flocking. *J. Theor. Biol.* **38**: 419–422. 6.

Rainey, R. H. (1967). Natural displacement of pollution from the Great Lakes. *Science* **155**: 1242–1243. 8.1.

*Rapoport, A. (1976). Directions in mathematical psychology. I. *Amer. Math. Mon.* **83**: 85–106. 2.1.

Rashevsky, N. (1960). *Mathematical Biophysics.* Vol. 2 (3rd ed.). Dover. 2.1.

Rashevsky, N. (1964). *Some Medical Aspects of Mathematical Biology.* C. C. Thomas. 9.2.

Richardson, L. F. (1960). *Arms and Insecurity.* N. Rashevsky and E. Trucco, eds. Boxwood Press and Quadrangle Books. See also, Mathematics of war and foreign politics. In J. R. Newman (1956, pp. 1240–1253). 9.2.

Riggs, D. S. (1963). *The Mathematical Approach to Physiological Problems.* Williams and Wilkins. 8.1.

*Roberts, F. S. (1976). *Discrete Mathematical Models.* Prentice-Hall. 5.2.

*Rosen, R. (1967). *Optimality Principles in Biology.* Plenum. 4.1.

Rosenzweig, M. L. (1971). Paradox of enrichment: Destabilization of exploitation ecosystems in ecological time. *Science* **171**: 385–387. This article resulted in interchanges between Rosenzweig and others. See *Science* **175**: 562–565; **177**: 902–904. 9.2.

Saaty, T. L. (1968). *Mathematical Models of Arms Control and Disarmament.* Wiley. 3.2, 5.1, 9.2.

Scheidegger, A. E. (1970). *Theoretical Geomorphology* (2nd ed.). Springer-Verlag. 5.2.

Schmidt-Nielsen, K. (1972). *How Animals Work.* Cambridge University Press. 2.1.

Schwartz, B. L. (1966). A new approach to stockout penalities. *Manage. Sci.* **12B**: 538–544. 4.1.

Schwartz, B. L. and M. A. B. Deakin (1973). Walking in the rain, reconsidered. *Math. Mag.* **46**: 272–276. 4.1.

Sedov, L. I. (1959). *Similarity and Dimensional Methods in Mechanics.* (Translated from the Russian 4th ed.). Academic Press. 2.2.

Shear, D. (1967). An analog of the Boltzmann H-theorem (a Liapunov function) for systems of coupled chemical reactions. *J. Theor. Biol.* **16**: 212–228. 9.2.

Simon, H. A. (1952). A formal theory of interaction in social groups. *Amer. Sociol. Rev.* **17**: 202–211. Reprinted in H. A. Simon (1957, pp. 99–114). 3.3.

Simon, H. A. (1955). On a class of skew distribution functions. *Biometrika* **42**: 425–440. Reprinted in H. A. Simon (1957, pp. 145–164). 10.

*Simon, H. A. (1957). *Models of Man Social and Rational.* Wiley.

Spector, W. S., ed. (1956). *Handbook of Biological Data.* Saunders. 2.1.

Sperry, K. (1967). Water and air pollution: Two reports on cleanup efforts. *Science* **158**: 351–355. 8.1.

Stahl, W. R. and J. Y. Gummerson (1967). Systematic allometry in five species of adult primates. *Growth* **31**: 21–24. 2.1.

Stevens, S. S. (1974). *Psychophysics.* G. Stevens, ed. Wiley. 2.1.

Swintosky, J. V. (1956). Illustrations and pharmaceutical interpretations of first order drug elimination rate from the bloodstream. *J. Amer. Pharm. Assoc.* **45**: 395–400. 8.2.

Synge, J. L. (1970). The problem of the thrown string. *Math. Gaz.* **54**: 250–260. 10.

Tanford, C. (1961). *Physical Chemistry of Macromolecules.* Wiley. 8.1.

Thom, R. (1975). *Structural Stability and Morphogenesis.* (Translated from the French.) Benjamin. 7.3.

Toomre, A. and J. Toomre (1972). Galactic bridges and tails. *Astrophys. J.* **178**: 623–666. 8.2.

Toomre, A. and J. Toomre (1973). Violent tides between galaxies. *Sci. Amer.* **229**(6): 38–48. 8.2.

Tsipis, K. (1975). Physics and calculus of countercity and counterforce nuclear attacks. *Science* **187**: 393–397. 5.1.

Tsipis, K. (1975a). The accuracy of strategic missiles. *Sci. Amer.* **233**(1): 14–23. 3.2.

Vidale, M. L. and H. B. Wolfe (1957). An operations-research study of sales response to advertising. *Oper. Res.* **5**: 370–381. Reprinted in R. L. Day and L. J. Parsons (1971, pp. 29–42) and in F. M. Bass et al. (1961, pp. 363–374). 8.1.

Vine, I. (1971). Risk of visual detection and pursuit by a predator and the selective advantage of flocking behavior. *J. Theor. Biol.* **30**: 405–422. 6.

Vold, M. J. (1959). A numerical approach to the problem of sediment volume. *J. Colloid Sci.* **14**: 168–174. 5.2.

Vold, M. J. (1959a). Sediment volume and structure in dispersions of anisometric particles. *J. Phys. Chem.* **63**: 1608–1612. 5.2.

Weihs, D. (1973). Mechanically efficient swimming techniques for fish with negative bouyancy. *J. Mar. Res.* **31**: 194–204. 4.1.

Wigner, E. P. (1960). The unreasonable effectiveness of mathematics in the natural sciences. *Comm. Pure. Appl. Math.* **13**: 1–14. 1.1.

Wilson, E. O. (1975). *Sociobiology.* Belknap. 4.2.

*Wilson, E. O. and W. H. Bossert (1971). *A Primer of Population Biology.* Sinauer Associates. 3.2, 4.2.

Woldenberg, M. J. (1969). Spatial order in fluvial systems: Horton's laws derived from mixed hexagonal hierarchies of drainage basin areas. *Geol. Soc. Amer. Bull.* **80**: 97–112. 5.2.

A GUIDE TO MODEL TOPICS

Models are grouped into major categories which are capitalized and grouped by affinity. Italicized numbers refer to chapters and sections that discuss a subject. Other numbers refer to problems dealing with the subject.

ASTRONOMY

colliding galaxies   8.1.2
number of comets   5.2.4

CHEMISTRY

chemical engineering   4.2.5
polymer formation   *8.1*, 8.1.3
reaction stability   9.2.8
sediment volume   *5.2*, 5.2.1

EARTH SCIENCES

particle sizes   *10*
reflected energy in the desert   *6*
sediment volume   *5.2*, 5.2.1
stream networks   *5.2*, 5.2.5
waves   2.2.4

PHYSICS

ENGINEERING

TRAFFIC

PSYCHOLOGY AND PSYCHOPHYSICS

HUMAN PHYSIOLOGY AND MEDICINE

ECONOMICS OF A FIRM

OTHER ECONOMICS

UNIVERSITIES

MISCELLANEOUS

INDEX

A CATALOG OF SELECTED
DOVER BOOKS
IN SCIENCE AND MATHEMATICS

Mathematics

FUNCTIONAL ANALYSIS (Second Corrected Edition), George Bachman and Lawrence Narici. Excellent treatment of subject geared toward students with background in linear algebra, advanced calculus, physics, and engineering. Text covers introduction to inner-product spaces, normed, metric spaces, and topological spaces; complete orthonormal sets, the Hahn-Banach Theorem and its consequences, and many other related subjects. 1966 ed. 544pp. 6⅛ x 9¼. 40251-7

ASYMPTOTIC EXPANSIONS OF INTEGRALS, Norman Bleistein & Richard A. Handelsman. Best introduction to important field with applications in a variety of scientific disciplines. New preface. Problems. Diagrams. Tables. Bibliography. Index. 448pp. 5⅜ x 8½. 65082-0

VECTOR AND TENSOR ANALYSIS WITH APPLICATIONS, A. I. Borisenko and I. E. Tarapov. Concise introduction. Worked-out problems, solutions, exercises. 257pp. 5⅝ x 8¼. 63833-2

THE ABSOLUTE DIFFERENTIAL CALCULUS (CALCULUS OF TENSORS), Tullio Levi-Civita. Great 20th-century mathematician's classic work on material necessary for mathematical grasp of theory of relativity. 452pp. 5⅝ x 8¼. 63401-9

AN INTRODUCTION TO ORDINARY DIFFERENTIAL EQUATIONS, Earl A. Coddington. A thorough and systematic first course in elementary differential equations for undergraduates in mathematics and science, with many exercises and problems (with answers). Index. 304pp. 5⅜ x 8½. 65942-9

FOURIER SERIES AND ORTHOGONAL FUNCTIONS, Harry F. Davis. An incisive text combining theory and practical example to introduce Fourier series, orthogonal functions and applications of the Fourier method to boundary-value problems. 570 exercises. Answers and notes. 416pp. 5⅜ x 8½. 65973-9

COMPUTABILITY AND UNSOLVABILITY, Martin Davis. Classic graduate-level introduction to theory of computability, usually referred to as theory of recurrent functions. New preface and appendix. 288pp. 5⅜ x 8½. 61471-9

ASYMPTOTIC METHODS IN ANALYSIS, N. G. de Bruijn. An inexpensive, comprehensive guide to asymptotic methods–the pioneering work that teaches by explaining worked examples in detail. Index. 224pp. 5⅜ x 8½ 64221-6

APPLIED COMPLEX VARIABLES, John W. Dettman. Step-by-step coverage of fundamentals of analytic function theory–plus lucid exposition of five important applications: Potential Theory; Ordinary Differential Equations; Fourier Transforms; Laplace Transforms; Asymptotic Expansions. 66 figures. Exercises at chapter ends. 512pp. 5⅜ x 8½. 64670-X

INTRODUCTION TO LINEAR ALGEBRA AND DIFFERENTIAL EQUATIONS, John W. Dettman. Excellent text covers complex numbers, determinants, orthonormal bases, Laplace transforms, much more. Exercises with solutions. Undergraduate level. 416pp. 5⅜ x 8½. 65191-6

TENSOR CALCULUS, J.L. Synge and A. Schild. Widely used introductory text covers spaces and tensors, basic operations in Riemannian space, non-Riemannian spaces, etc. 324pp. 5⅜ x 8¼.   63612-7

ORDINARY DIFFERENTIAL EQUATIONS, Morris Tenenbaum and Harry Pollard. Exhaustive survey of ordinary differential equations for undergraduates in mathematics, engineering, science. Thorough analysis of theorems. Diagrams. Bibliography. Index. 818pp. 5⅜ x 8½.   64940-7

INTEGRAL EQUATIONS, F. G. Tricomi. Authoritative, well-written treatment of extremely useful mathematical tool with wide applications. Volterra Equations, Fredholm Equations, much more. Advanced undergraduate to graduate level. Exercises. Bibliography. 238pp. 5⅜ x 8½.   64828-1

FOURIER SERIES, Georgi P. Tolstov. Translated by Richard A. Silverman. A valuable addition to the literature on the subject, moving clearly from subject to subject and theorem to theorem. 107 problems, answers. 336pp. 5⅜ x 8½.   63317-9

INTRODUCTION TO MATHEMATICAL THINKING, Friedrich Waismann. Examinations of arithmetic, geometry, and theory of integers; rational and natural numbers; complete induction; limit and point of accumulation; remarkable curves; complex and hypercomplex numbers, more. 1959 ed. 27 figures. xii+260pp. 5⅜ x 8½. 42804-4

POPULAR LECTURES ON MATHEMATICAL LOGIC, Hao Wang. Noted logician's lucid treatment of historical developments, set theory, model theory, recursion theory and constructivism, proof theory, more. 3 appendixes. Bibliography. 1981 ed. ix+283pp. 5⅜ x 8½.   67632-3

CALCULUS OF VARIATIONS, Robert Weinstock. Basic introduction covering isoperimetric problems, theory of elasticity, quantum mechanics, electrostatics, etc. Exercises throughout. 326pp. 5⅜ x 8½.   63069-2

THE CONTINUUM: A Critical Examination of the Foundation of Analysis, Hermann Weyl. Classic of 20th-century foundational research deals with the conceptual problem posed by the continuum. 156pp. 5⅜ x 8½.   67982-9

CHALLENGING MATHEMATICAL PROBLEMS WITH ELEMENTARY SOLUTIONS, A. M. Yaglom and I. M. Yaglom. Over 170 challenging problems on probability theory, combinatorial analysis, points and lines, topology, convex polygons, many other topics. Solutions. Total of 445pp. 5⅜ x 8½. Two-vol. set.
                                             Vol. I: 65536-9   Vol. II: 65537-7

Paperbound unless otherwise indicated. Available at your book dealer, online at **www.doverpublications.com**, or by writing to Dept. GI, Dover Publications, Inc., 31 East 2nd Street, Mineola, NY 11501. For current price information or for free catalogs (please indicate field of interest), write to Dover Publications or log on to **www.doverpublications.com** and see every Dover book in print. Dover publishes more than 500 books each year on science, elementary and advanced mathematics, biology, music, art, literary history, social sciences, and other areas.